IMAGES OF WAR

M65 ATOMIC CANNON

RARE PHOTOGRAPHS FROM WARTIME ARCHIVES

David Doyle

Pen & Sword
MILITARY

First published in Great Britain in 2019 by
PEN & SWORD MILITARY
An imprint of
Pen & Sword Books Ltd
47 Church Street
Barnsley
South Yorkshire
S70 2AS

ISBN 978-1-52674-360-2

A CIP catalogue record for this book is available from the British Library.

Typeset by Concept, Huddersfield, West Yorkshire HD4 5JL
Printed and bound in China by Printworks Global Ltd

Pen & Sword Books Limited incorporates the imprints of Atlas, Archaeology, Aviation, Discovery, Family History, Fiction, History, Maritime, Military, Military Classics, Politics, Select, Transport, True Crime, Air World, Frontline Publishing, Leo Cooper, Remember When, Seaforth Publishing, The Praetorian Press, Wharncliffe Local History, Wharncliffe Transport, Wharncliffe True Crime and White Owl.

For a complete list of Pen & Sword titles please contact
PEN & SWORD BOOKS LIMITED
47 Church Street, Barnsley, South Yorkshire S70 2AS, England
E-mail: enquiries@pen-and-sword.co.uk
Website: www.pen-and-sword.co.uk

Contents

Acknowledgements

This book would not have been possible without the generous help of many friends and colleagues. I would especially like to thank Tom Kailbourn, Scott Taylor, Tom Jett, Gordon Blaker at the Artillery Museum, Kurt Laughlin, the staff of the National Archives, Dana Bell, the late Joe Ramsdell of Pacific Car and Foundry (PACCAR), the TACOM LCMC History Office, the Rock Island Arsenal Museum, and Jim Gilmore. I am especially indebted to my wife Denise, who not only scanned hundreds of documents and photos for this project, but every day shows me what a blessing our marriage is.

Introduction

Since the advent of artillery, there has seemingly been the desire by strategists for longer range or heavier hitting field pieces. Such was the case in 1945 when the War Department Equipment Board (the Stillwell Board), addressing the future needs of the US Army, remarked 'with expected improvements, guided missiles and all-weather bombing should meet this requirement. In the event that such expectations are not realized, large caliber, long range howitzers and guns [i.e. than the 8-inch gun M1 and Howitzer M1, now standard] may be required.'

The 1945 recommendation, published in 1946, dovetailed nicely into work begun in 1944. That effort, performed by the Artillery Development Division Sub-Office, Franklin Institute, Philadelphia, and approved by the Ordnance Technical Committee (OCM) item nos 25587 and 25775, led to the Gun, 240mm, T1, and Carriage, Gun, 240mm, T2. Pilot models of the gun and carriage were placed on order, and fabrication begun. The pilot gun was produced in 1948 by Watertown Arsenal, Massachusetts.

Initially, Limber, Heavy, Carriage, T10; Wagon, Transport, Carriage, T20, and Wagon, Transport, Carriage, T19, were planned, but OCM 32732 of February 1949 recommended that Tank Transporter T8E1 be modified to transport the Gun, 240mm, T1 instead. The previously planned limber and carriages never advanced beyond the drawing boards.

The aforementioned Tank Transporter T8E1 was the most recent development in a succession of experimental double-end transporters. In July 1942 OCM 18517 had initiated the study of the 'Tank Recovery Vehicle, T1' and arrangements had been made with Cook Brothers (Allied Machinery Manufacturing Company) of Los Angeles to submit engineering studies. The planned vehicle, described as an 8 × 8, featured independently powered four-wheel bogies at each end of a platform for carrying a tank. Two pilot vehicles were ordered for testing, over the concerns of the Corps of Engineers, who objected to the load limits of the vehicle, and the Army Ground Forces. The latter objected due to the projected considerable weight and size, and also because of the considerable promise held by a similar vehicle under construction by Le Tourneau, the T4. Accordingly, the Army Service Forces cancelled the T1 project on 3 September 1943, before either pilot was completed.

The Le Tourneau T4, like the aborted T1, was a double-ended vehicle with a load platform suspended between the power units. The T4 design differed from the planned T1 in that while each power unit of the T4 had only a pair of wheels (rather than the four of the T1), each power unit also boasted a pair of Cadillac V8 engines. It should be noted that unlike the T1, the T4 was a true double-end design, with operator controls at each end, and capable of operating equally well in either direction of travel.

Work on this vehicle was initiated by OCM 19265 of 3 December 1942. A contract in the amount of $39,500, for the design and fabrication of one pilot, was issued 6 January 1943 and the pilot was completed in August 1943.

Initially planned for testing at Camp Seeley, California, the closure of that facility caused the T4 to be re-routed to Aberdeen Proving Ground. Initial engineering tests of the T4 were completed in July 1944 at Aberdeen. It was noted that the off-road performance of the vehicle was exceptional, but that loading tanks, especially under simulated combat conditions, was difficult and time-consuming. Further, the four-engine and transmission powertrain proved maintenance intensive.

The next step in the development of the double-end tank transporter was the Mack T8. The Philadelphia Ordnance District issued contract W36-034-ORD-1608 for the engineering of this vehicle in March 1944, followed six months later with a contract for production of a pilot. By March 1945 that contract had been upped to include three pilots. The T8 had four wheels per power unit, as opposed to the two wheels of the T4, and was not a true double-ender as there were drivers in both units, and both drivers faced the same direction.

A 1,090 cubic inch displacement Hall-Scott model 441 six-cylinder engine powered each of the two units, driving through Spicer 3F1R torque-converter transmissions. The design permitted the rear power unit to be readily uncoupled from the cargo platform, allowing for quick, easy loading and unloading.

The T8E1 followed the same lines as the T8, but each unit was powered by a Ford GAA tank engine in lieu of the Hall-Scott powerplant, which effectively doubled the available horsepower in the trucks.

Three prototypes of the T8 and T8E1 were produced, which underwent extensive testing at Aberdeen in 1945-46.

With the conclusion of the war against Japan and the army's bulging roster of vehicles thereafter, no production of the T8 or T8E1 was pursued. The vehicles languished until 1949, when interest in the double-end concept was renewed as a means to transport the T1 240mm cannon.

Modifications to the T8E1 resulted in the T9 Transporter. While the T8 had an impressive 59 foot, 4.5 inch length, the T9, with the T1 gun and T2 carriage supported, stretched out to 71 feet.

The Atomic Cannon

By 1949 the army wanted artillery that could deliver a nuclear warhead. The means to do this fell to Robert M. Schwartz, a civilian employee of Picatinny Arsenal, detailed to the Pentagon. Schwartz scaled up the 240mm shell to 280mm, and improved the design so it would withstand 6,000 revolutions per minute. This required that the shell be 4,000 times stronger than the casing for an atom bomb.

In addition to the nuclear round, which initially housed a W9 warhead, conventional high explosive rounds were developed.

The carriage design of the previously developed 240mm Gun T1 required only slight modifications to accommodate the 280mm barrel. The pilot 280mm carriage was produced at Watertown Arsenal, while the barrel breech for the pilot (as well as subsequent production units) was produced at the army's Watervliet Arsenal, Watervliet, New York.

Dravo Corporation, a Pittsburgh-based firm, along with Baldwin Locomotive Works, produced the T72 production carriages while the recoil mechanisms were produced by R. Hoe and Company, a New York-based printing press manufacturer.

The first 280mm version, initially designated the T131 and later standardized as the M65, was completed in 1951. In October 1952, three of the massive cannons and their transporters were the centerpiece of a public demonstration at Aberdeen Proving Ground, Maryland.

After many successful test firings with conventional 280mm ammunition, it was time to test the gun in the role for which it was intended, firing the T124 projectile with W-9 15 kiloton nuclear warhead.

Two guns, assigned to the 867th Field Artillery Battalion, Fort Sill, Oklahoma, were transported by rail to Nellis Air Force Base, near Las Vegas. The guns, coupled to their prime movers and joined by their support vehicles, then convoyed to Frenchman Flat. After both guns fired conventional munitions to evaluate atmospheric conditions on 24 May 1953, at 0830 the next day a battery gun 'Able Annie' let loose with the only nuclear round the type would ever fire.

This test, known as Shot Grable, were part of a series of tests known as Upshot-Knothole, and Shot Grable revealed that a relatively low yield atomic weapon detonated at a low altitude produced considerably more damage, kiloton for kiloton, than a higher yield weapon detonated at either higher altitudes or at ground level.

The 867th, which made this shot, was one of six Field Artillery Battalions, 280mm Gun that were raised in 1952. Five of the units, the 59th, 264th, 265th, 858th, and 867th were raised at Fort Sill, while the 663rd was raised at Fort Bragg.

The battalions raised at Fort Sill were ultimately assigned to the 7th Army in Europe as part of the 42nd Field Artillery Group. The units began arriving in Germany in October 1953. The 663rd was sent to Okinawa in July 1955 as a deterrent to further aggression in Korea.

Yet another 280mm Battalion was raised at Fort Sill, the 216th Field Artillery Battalion. This unit too was deployed to Europe, arriving in the spring of 1955.

The 613rd Field Artillery Battalion (280mm), originally based at Fort Bragg, as part of the VIII Airborne Corps Artillery arrived in March 1957 replacing the 59th Field Artillery Battalion as part of Operation Gyroscope. This was an exchange of personnel, rather than actual hardware.

The 280mm gun battalions, as one would imagine, were intended as a counter to the threat of a Soviet overrun of Germany – and potentially all of Europe. Although few in number, the weapons, while effective deterrents, presented several logistical challenges.

Even with the extensive flexibility afforded by the design, negotiating narrow European streets was a challenge, and several buildings were damaged by too-tight turns during the M65s deployment.

The great weight, even though distributed over twelve very large tires, caused roadway damage and collapse, especially if driven too close to the shoulder of the highway. More than one cannon and transporter combo rolled over during such situations, with considerable damage and loss of life.

The weapons were maintenance-intensive, even without the need for wreck-caused repairs. In fact, late 1955, General Hodes ordered the 42nd Field Artillery Group to conduct a one-time school to cross-train mechanics.

These issues, as well as the limits of the range of the weapon, led to the development and deployment of the 'Honest John' 762mm rocket to augment the M65 as a nuclear deterrent. The new weapons, mounted on 5-ton 6 x 6 chassis, began to arrive in Europe in 1954, and a year later an atomic munition was developed for the 8-inch gun, a weapon which was abundant in Europe.

Although more mobile and possessing longer range, these weapons were notably less accurate than the M65. Time would bring about improved accuracy of the rocket artillery, which along with the availability of nuclear weapons in smaller calibers, led to other systems augmenting and then supplanting the M65. It remained in the American arsenal until November 1963.

An atomic cannon and, on the low end of US Army field artillery pieces, a 75mm pack howitzer, were photographed together at Fort Sill in August 1961 to emphasize the massive size of the 280mm gun M65. The light-colored spot on the side of the 280mm gun breech is artwork showing a nuclear mushroom cloud. *(NARA)*

The development of both the 240mm gun T1 and the 280mm atomic cannon T131/M65, very heavy, mobile artillery pieces, was dependent on the concurrent development of suitable artillery prime movers: in these cases, front and rear 2 × 2 tractors that supported the gun and its carriage. An early step in the process was the LeTourneau T4 tank transporter, produced and tested in 1944 and seen here from the right rear carrying an M4 medium tank on a platform connected to the two tractors. *(TACOM LCMC History Office)*

As seen from the right front, the front tractor of the LeTourneau T4 tank transporter had two separate, enclosed cabs, as well as a winch-operated dozer blade. The tractors and the platform constituted a unit that could not be readily disassembled, and it was necessary to load the tank to be transported by using winch-power to pull it sideways onto the transporter. *(TACOM LCMC History Office)*

The LeTourneau T4 tank transporter is viewed from the right side with an empty platform between the two tractors. The photo was taken at Aberdeen Proving Ground on 8 April 1944. In this image, the armored top panels of the doors are in the raised or closed positions. *(NARA)*

The T4 tank transporter relied on a system of loading dollies and channels along with a winch, sheaves, and cables, to load a tank sideways onto the platform: a time-consuming process. The dollies and channels are shown assembled for loading a tank in a 17 July 1944 Aberdeen Proving Ground photo. *(NARA)*

(**Above**) While crossing a bridge in Perryman, Maryland, on 17 May 1944, the T4 tank transporter has become mired in a roadside ditch. The view is from the front of the forward tractor. Details are visible of the two separate cabs and the dozer blade. A detachable-type headlight assembly and brush guard were on the inboard side of each cab. A spotlight and a siren were on the left cab roof. *(NARA)*

(**Opposite, above**) The T4 tank transporter is viewed from the left rear after becoming stuck in a ditch at Perryman, Maryland, on 17 May 1944. Each of the tractors of the T4 units had two Cadillac 90-degree V8 engines, with a total horsepower of 440 at 3400 rpm. *(NARA)*

(**Opposite, below**) As viewed from still another angle, the mired T4 tank transporter was carrying a pilot T22 medium tank on the platform. A clear view is offered of the rear tractor's cab and machine gun tub. A machine gun carriage and cradle are visible atop the gun tub. *(NARA)*

(**Above**) In July 1945, during the year after LeTourneau produced its T4 tank transporter, Mack delivered to the army a pilot heavy tank transporter, the T8. Its concept was two 4 × 4 tractors with a platform between them. To load a tank on the transporter, the rear tractor was detached and the platform lowered so the tank could be driven or winched aboard. *(Tony Gibbs)*

(**Opposite, above**) A GI stands next to the rear tractor of the Mack T8 heavy tank transporter, giving a sense of the massive scale of the vehicle. The cargo platform had a gooseneck at each end that pivoted on a tractor. An identification feature for differentiating the front from the rear gooseneck is that the rear one had smooth sides, while the front one had recessed sides. *(TACOM LCMC History Office)*

(**Opposite, below**) An 8 January 1946 Aberdeen Proving Ground photograph shows the interior of the cab of the forward tractor of a Mack T8 heavy tank transporter, as seen from the assistant driver's seat. On the front of the steering column is the shifting lever for the automatic torque-converter transmission. *(NARA)*

T8-645V8398.

(**Opposite, above**) An M26 Pershing tank is loaded on the Mack T8 heavy tank transporter, with the front tractor to the right. To load or unload a tank, the rear end of the platform was lowered by hydraulic power and detached from the rear tractor, so that the platform now formed a ramp. Then, the rear tractor was reconnected and the rear of the platform was raised and secured. *(NARA)*

(**Opposite, below**) The front tractor and the forward gooseneck of a Mack T8 heavy transporter are depicted. At the front end of the gooseneck was a Garwood winch assembly. The tires of the tractor were Firestone size 21.00-28. The cab had front and rear windshields with windshield wipers. A ladder attached to the running board and the front bumper provided access to the cab. *(Tony Gibbs)*

(**Above**) The Mack T8 heavy tank transporter is seen from above and to the right in a 7 January 1946 photo from Aberdeen Proving Ground. The forward tractor is to the left. The ramps of the transporter platform were turned down slightly at the rear so that boarding vehicles could more easily mount the ramp. *(NARA)*

(**Below**) The rear tractor has been detached from a T8 heavy tank transporter and has been stabilized in a manner to enable it to be driven away. The platform of the transporter is in the lowered position, ready for a tank or other vehicle to be boarded. The photo was taken on 8 January 1946 at Aberdeen. *(NARA)*

In a view of a Mack T8 with stakes installed on the cargo platform, details of the front of the forward tractor are visible. At the center of the cab is a grille protecting the radiator. On the lower right side of the grille is a pioneer tool rack. Two canvas doors with soft-plastic windows were on each side of the cab. In addition to the T8 prototype, Mack received a contract to build three pilot T8E1 heavy tank transporters. The army tested the T8 and the T8E1s in 1945. *(Tony Gibbs)*

Although the T8 heavy tank transporter program was discontinued after the prototype and pilot vehicles were tested, the two tractors of one of the T8E1 pilots were modified for use as prime movers for the first atomic cannon, a T131 280mm gun on the T2 240mm gun carriage. That combination is seen from the right side, with the forward, or puller tractor to the right and the rear, pusher tractor to the left. Note the large exhaust stack on each tractor, not present on the T8 tractors. *(Rock Island Arsenal Museum)*

Chapter One

T1 in Detail

Twenty examples of the atomic cannon were manufactured. There are several surviving examples, including this one preserved at the Virginia War Museum, Newport News, Virginia. The atomic cannon constituted the 280mm Gun T131 (standardized as the M65 in 1960) mounted on the 280mm Gun Carriage T72 (standardized as the M30). The main part of the gun carriage, which, in its firing position, is elevated several feet above the ground, is called the top carriage. *(Tom Jett)*

(**Opposite, above**) Below the rear of the top carriage is the float: the rear structure on which the gun rested when deployed for firing. The float incorporated a mechanism for traversing the gun 7.5 degrees to either side of center. The entire traverse of the gun could be realigned by lifting the rear of the top carriage, including the float, and repositioning it. *(Tom Jett)*

(**Opposite, below**) The atomic cannon is viewed from the right rear. On bottom corners of the rear of the top carriage are lifting hooks, which engage with shackles on the hoist cylinders on the lifting forks to raise and lower the cannon and carriage. Also on the rear of the carriage is an access door. Two tubular tie rods are attached to the bolsters of the firing base and the float; the left tie rod is visible directly below the top carriage. *(Author)*

(**Above**) The breech of the atomic cannon is seen from the rear. On the rear face of the breech, near the top is a horizontally oriented eye, for attaching a winch cable to pull the gun tube from its travel position into its firing position. Below it is the round breech block. In the foreground is part of the loading trough, on which ammunition was placed prior to being rammed into the breech. *(Tom Jett)*

Elements along the left side of the top carriage include, left to right: three folding platforms in the lowered positions; the ammunition-loading davit; the case for the manually operated davit winch, with a lug for a hand crank towards the bottom; and an electrical box. *(Author)*

The left side of the atomic cannon is depicted. The bottom edge of the top carriage formed a slide for secondary recoil purposes; note the ribbed guides for this purpose adjacent to the top of the leveling jack with caster. Such guides are on the sides of the bolsters of the bottom carriage, specifically, the firing base (front) and the float (rear). *(Author)*

The following series of photographs, documenting the construction of atomic cannons, was taken at Baldwin Locomotive Works plant at Eddystone, Pennsylvania, on 24 September 1952. Here the top carriage of an atomic cannon is under construction. The view is from the rear of the assembly. *(Baldwin Locomotive Works)*

9-3-52
14310-18

(**Opposite, above**) Work is proceeding on a cradle assembly, which is secured in an upside-down position, with the cradle-elevating rack, with its large cogs, on top in the background. The workers are checking the machining of the cogs using a special rack fitted with gauges. Under the rags on the cradle in the foreground is a socket for locking the cradle to the top carriage during travel; a similar socket was on the opposite side of the cradle. *(Baldwin Locomotive Works)*

(**Opposite, below**) The installation of the rear of the top carriage onto the float is viewed from the left side. The bottom carriage, which supported the top carriage and the gun when emplaced on the ground, constituted the float, to the rear, and the firing base, to the front. The float was equipped with a traversing rack, by means of which the entire carriage and gun could be traversed a total of 15 degrees. *(Baldwin Locomotive Works)*

(**Above**) On 24 September 1952, an atomic cannon carriage is under construction to the left at the Baldwin Locomotive Works, while the 280mm gun and cradle have been installed on the carriage at the center of the photo. The loading trough, on which ammunition would be placed for ramming into the gun, is visible to the rear of the breech. *(Baldwin Locomotive Works)*

9-24-52

(**Opposite, above**) The same atomic cannon seen at the center of the preceding photo is viewed from the factory floor. The object standing up to the rear of the 280mm cannon breech is the loading trough, in the folded position. Electrical conduits on the side of the carriage show up as light-colored lines. *(Baldwin Locomotive Works)*

(**Opposite, below**) The same atomic cannon under construction is viewed from an elevated position to the left side. The cannon, cradle, carriage, and most accessories have been painted, but the light-colored section of barrel to the front of the cradle was not painted, nor would it ever be. It was left in a polished-steel finish because it made contact with the interior of the cradle during recoil and recuperation. *(Baldwin Locomotive Works)*

(**Above**) An audience made up exclusively of men in suits is gathered next to a newly minted atomic cannon at the Baldwin Locomotive Works factory in Eddystone, Pennsylvania, on 10 March 1953, apparently as part of a completion ceremony for the piece. The polished, bare-metal portion of the 280mm cannon barrel to the front of the cradle shows up to particularly good advantage. *(Baldwin Locomotive Works)*

Chapter Two

T10 Development

The third and final link in the evolutionary chain of front-and-rear-tractor transporters that made possible a transporter for the atomic cannon, following the T4 and T8 tank transporters, was the Heavy Artillery Transporter T10. Produced by the Kenworth Motor Truck Company, this transporter consisted of a front truck, designated T10A (seen to the right) and a rear one, T10B, between which was suspended the atomic cannon and its carriage. Although the designations T10A and T10B would be changed to M249 and M259 by 1955, the combined transporter system would continue to be called the T10. A T10 transporter is seen here undergoing load tests using a special pallet with extra weights on top of it. *(PACCAR)*

In a photograph dated 7 October 1952, a T10 transporter is being subjected to load tests. To the left is the 'B' truck, with yellow and black caution stripes on the rear bumper and US Army registration number (or USA number) 60146082 on the rear and the side door. The test pallet was latched to lifting forks at the rear of the A truck and the front of the B truck, and these forks would serve to hold the carriage of the atomic cannon when it was being transported. *(PACCAR)*

A different T10 heavy artillery transporter with a test pallet is viewed from the right side in an Aberdeen Proving Ground photograph dated 7 February 1952. The front, A, truck is USA number 60147222, and the B truck is 60147223. The two trucks were significantly different in design, with the front unit easily identifiable by its spare tire. Both trucks had the engine compartment to the rear of the cab; the front truck's cab jutted to the front of the front tires, while the rear truck's engine compartment protruded far aft of the rear tires. Only the A truck was equipped with a winch and a spare tire. *(TACOM LCMC History Office)*

(**Above**) A T10 heavy artillery transport is seen from the right side on 9 June 1953. To the right is the forward truck, USA number 60146771, and to the left is the rear truck, 60146772. The lifting forks of the two trucks were of different shapes and dimensions, the one of the forward truck being shorter in length and having a more pronounced bend at the bottom than that of the rear truck. *(TACOM LCMC History Office)*

(**Opposite, above**) T10A heavy artillery truck USA number 60147222 is the subject of a 7 February 1952 Aberdeen Proving Ground photo. These trucks were equipped with size 16.00-25 tires. They were powered by Ordnance AO895-4 engines through Ordnance torque-converter transmissions. *(NARA via Jim Gilmore)*

(**Opposite, below**) The same T10A truck. On both the A and the B truck, the lifting forks were attached to fifth wheels permanently mounted on the chassis, to allow for horizontal and vertical swiveling. The lower parts of the hydraulic hoist cylinders are visible under the lifting fork. Behind the engine compartment are the winch, with a spotlight on a pedestal mount, and the spare tire. Two windows are on the rear of the cab. *(NARA via Jim Gilmore)*

U.S.A. 60147222

U.S.A. 60147222

The front and the rear of the cab of T10A USA number 60147222 is shown with a canvas top attached in an October 1952 photo. In the top photo is the davit with a hand-operated winch for raising and lowering the spare tire. In the lower photo, on the deck to the rear of the cab, is the fan-shroud shield. The design of this feature would change later, with the addition of a raised muffler shield at the center of the fan-shroud shield. *(NARA via Jim Gilmore)*

The interior of the cab of T10 truck USA number 60146753 is seen through the left door in an October 1952 photo. On the interior of the door is a grab handle. Inside are the driver's adjustable seat, the steering wheel and steering column, and the instrument panel. On the right rear corner of the cab are intercom headphones. The intercom system was essential for coordinating the movements and functions of the front and the rear trucks.
(NARA via Jim Gilmore)

A 1 September 1954 Aberdeen Proving Ground photograph shows the interior of the cab of an XM249. It was taken to document the proposed location of two carbines below the rear windows of the cab. The bench seats in the cab were provided with four sets of cushions. Note the grab handle on the edge of the roof.
(NARA via Jim Gilmore)

The atomic cannon would be vulnerable to aerial attack, so efforts were made to provide its transports with anti-aircraft machine guns. T10A USA number 60147222 was armed with a .50-caliber machine gun on a ring mount above the right side of the cab roof, as documented in a 28 May 1952 photograph at Aberdeen Proving Ground. *(NARA via Jim Gilmore)*

As initially configured, the T10A trucks were 126.5 inches tall with the spare tire installed. This height was excessive for purposes of railroad shipment, so measures were taken in 1954 to shorten the vehicle by removing the cab roof, the spare tire and its davit, and the spotlight. These changes resulted in an overall height of 112.5 inches. By the time this photo was taken on 2 July 1954, the T10A had been redesignated the XM249. By early 1955, test reports would be referring to it and its rear mate as the M249 and M250. *(NARA via Jim Gilmore)*

This T10A truck is viewed from above, showing the position of the ring mount on the right side of the cab roof. The ring was supported by three brackets attached to the roof. Clear details are also visible of the top of the engine compartment, the muffler and its perforated shield, the winch, the fifth wheel, and the lifting fork. *(NARA via Jim Gilmore)*

T10B rear truck USA number 60147223 is viewed from the left front on 7 February 1952. The big shackles on the bottoms of the hydraulic hoisting cylinders on the lifting fork are stored for travel on hooks to the fronts of the elliptical springs of the front suspension. On the bottom of the fork slightly forward of the hoisting cylinders are fittings for locking the gun carriage in place by means of steel wedges. On the side of the fork are the brake override service line, the engine-throttle override service line, and shutoff valves. *(NARA via Jim Gilmore)*

The same T10B is viewed from the right rear, showing the engine compartment to the rear of the cab, with its screen-type ventilation grilles on the side. A holder for a 5-gallon liquid container is on the left rear of the compartment. On top of the engine compartment is the muffler, which sits inside the fan shroud extension. The design of the muffler and fan shroud would change significantly by the late 1950s. *(NARA via Jim Gilmore)*

The same XM250B with the machine gun ring mount is viewed from over the lifting fork on 28 May 1952. A canvas cover to keep out the elements is fitted to the opening on the cab roof, and a dust cover is over the machine gun and the ammunition box. *(NARA via Jim Gilmore)*

U.S.A.
60147223
U.S.ARMY
ORDNANCE

U.S.A.60147223

(**Opposite, above**) An XM250B rear truck, USA number 60147223, is shown during testing at Aberdeen Proving Ground on 1 July 1954 after the top of the cab had been removed to lower the height of the vehicle for purposes of rail shipment. The reduced height was 120 inches, as opposed to 137.5 inches before the top was removed. *(NARA via Jim Gilmore)*

(**Opposite, below**) At Aberdeen Proving Ground on 28 May 1952, XM250B USA number 60147223 has been fitted with a ring mount for a .50-caliber machine gun on the right side of the cab roof. The lifting fork is in the travel position when not transporting a gun, with the shackles of the hoisting cylinders attached to hooks on the truck chassis. *(NARA via Jim Gilmore)*

(**Above**) A T10 heavy artillery transporter is being subjected to flexion tests with a test pallet coupled to the two trucks. In the foreground is the A truck on a 10-degree side slope, while the B truck in the background is on level ground. Both truck cabs are equipped with ring-mounted .50-caliber machine guns. Each of the weighs on top of the pallet has a lifting eye on top. *(NARA via Jim Gilmore)*

During weight trials at Aberdeen Proving Ground on 6 March 1952, a T10 heavy artillery transporter has become immobilized on a cross-country course. In its attempt to extricate itself, the front truck, foreground, has sprayed the lifting fork with mud. *(NARA via Jim Gilmore)*

A T10 heavy artillery transporter with a test pallet hitched between the front truck, USA number 60147222, right, and the rear one, USA number 60147223, negotiates water that is 6 feet deep in the fording basin at Aberdeen Proving Ground on 10 June 1953. 'U.S. ARMY/ORDNANCE' is stenciled above the USA number on the cab door of the front truck. 'TEST LOAD' is marked on the side of the test pallet. During fording tests it was found that water seeped into the axles and the toolboxes. *(NARA via Jim Gilmore)*

Chapter Three

M65 Development and Testing

A Heavy Artillery Transporter T10 coupled to a 280mm Gun T131 pauses during tests at Aberdeen Proving Ground in September 1952. The front tractor, T10A US Army registration number 601467749, is in the foreground. *(NARA)*

(**Above**) The T10 has paused while it executes a sharp right turn with the 280mm Gun T131 during trials at Aberdeen Proving Ground in September 1952. The front tractor, registration number 60146749, is in the foreground, offering a clear view of the side of the cab, the engine housing, and the winch. *(NARA)*

(**Opposite, above**) The T10A front tractor is seen from the front during a pause at Aberdeen in September 1952. A service headlight and a marker light were in a teardrop-shaped recess on each side of the front of the vehicle. In another recess between those lights was a warning light. *(NARA)*

(**Opposite, below**) The rear-end teammate of T10A registration number 60146749, T10B 60146750, presents its rear to the camera as the vehicles execute a left turn at Aberdeen. Atop the rear deck of this vehicle is the muffler, positioned inside the fan shroud extension. The army's practice was to assign a registration number of the T10B that was one higher than that of the T10A front tractor. *(NARA)*

U.S. ARMY
ORDNANCE

U.S.A. 60146750
U.S. ARMY
ORDNANCE

U.S. ARMY
ORDNANCE

U.S. ARMY
ORDNANCE

(**Opposite, above**) A longer view of the T10 hauling the atomic cannon around a left curve is shown during trials at Aberdeen Proving Ground in September 1952. The drivers of the tractors were in communication with each other by interphone while transporting the cannon, as close coordination of the two tractors was necessary, especially during turns and similar maneuvers. *(NARA)*

(**Above**) The Heavy Artillery Transporter T10, comprising T10A 60146749 (foreground) and T10B 60146750, executes a sharp turn over a railroad track while carrying an atomic cannon during trials at Aberdeen in September 1952. *(NARA)*

(**Opposite, below**) Two T10 transporters laden with atomic cannons proceed towards the cameraman's position on a road at Aberdeen Proving Ground, September 1952. Suspended below the so-called top carriage, that is, the main carriage of the gun, is the bottom carriage, consisting of two main elements: a large turntable called the firing base, and a smaller rear support called the float. *(NARA)*

The T10 transporter comprising T10A 60146747 and T10B 60146748 carry the atomic cannon over a dip in the ground at Aberdeen Proving Ground. The carriage of the cannon had a ground clearance of 36 inches, which presumably was the space below the firing base and the ground on a level surface. *(NARA)*

The rear T10A tractor, registration number 60146748, negotiates a dip in the terrain while transporting an atomic cannon at Aberdeen. The rear end of the lifting fork was attached to the tractor's fifth wheel, while the front end of the fork was rigidly locked to the top carriage through a selection of steel wedges and sockets. *(NARA)*

In this 6 November 1952 photo at Aberdeen, the front tractor of 'A unit' of a T10 transporter hauls the barrel of an atomic cannon into battery position using a winch cable routed around a sheave attached to the top of the gun-cradle assembly and fastened to the breech of the gun. *(NARA)*

(**Above**) The T10B rear tractor of a Heavy Artillery Transporter T10 is viewed from the rear as the vehicle approaches the Bailey bridge at Aberdeen Proving Ground on 7 August 1953. The front end of the transporter has almost cleared the far end of the bridge.

(**Opposite, above**) At the 15 October 1952 press show, a T10A tractor crewman operates the winch in order to haul the barrel of the atomic cannon into battery position. The grilles on the side of the engine compartment to his rear are formed of wire screens. *(NARA)*

(**Opposite, below**) The atomic cannon and its T10 transporter were unveiled to the public at Aberdeen Proving Ground on 15 October 1952, in the first press show of that weapon. In this photo the gun crew is placing the piece into firing order. The T10B rear tractor is in the right background with its rear to the camera. *(NARA)*

(**Opposite, above**) The hand-cranked loading davit on the left side of the gun has been used to hoist a projectile for loading. Next it will be lowered onto the rammer trough and rammed into the chamber. The lever with a quadrant on the side of the top carriage below the projectile is the elevating speed-control handle. *(NARA)*

(**Opposite, below**) In another photograph from the 15 October 1953 press show at Aberdeen Proving Ground, a 280mm shell is on the trough and is being rammed into the breech. Once the projectile was laid on the trough, the whole ramming operation only took about three seconds. After the projectile is seated in the chamber, bagged propellant charges will be similarly rammed into the chamber. *(NARA)*

(**Above**) Several crewmen push on the rear portion of the top carriage of the atomic cannon during the press show at Aberdeen. The piece could be traversed a total of 15 degrees through a rack assembly on the float, and, because the mass of the piece was balanced on the firing plate, the gun could be traversed through 360 degrees by raising the float from the ground and having five crewmen push on the rear of the top carriage, thus swinging the entire piece around the axis of the firing base. *(NARA)*

U.S. ARMY
ORDNANCE
U.S. ARMY
ORDNANCE

(**Opposite, above**) The crew of an atomic cannon are preparing to transport the piece at Aberdeen in September 1952. The soldier on the ground is motioning the driver as he backs up the T10A front tractor. The blunt tips of the lifting fork will be positioned in sockets on the top carriage; the right rear socket is visible to the extreme right. Then the piece will be lifted and the lifting fork will be locked firmly to the top carriage. *(NARA)*

(**Opposite, below**) The tips of the lifting fork of the T10A front tractor have been maneuvered into the sockets on the top carriage, and a cable from the winch on the vehicle is retracting the gun barrel into its travel position. This T10A is US Army registration number 60146749. *(NARA)*

(**Above**) The 280mm gun barrel has been drawn back to its travel position, the tips of the lifting forks of both tractors have been seated in the sockets on the top carriage, and the hydraulic lifting hooks mounted on the lifting forks have been secured to the carriage. All appears to be in readiness for lifting the gun from the ground and locking the top carriage to the lifting forks. *(NARA)*

A Heavy Artillery Transporter T10 is transporting a 280mm Gun T131 'atomic cannon' across a Bailey bridge during a public demonstration of weapons and equipment at Aberdeen Proving Ground on 8 October 1953. In the foreground is the rear tractor, T10B registration number 60147680. *(NARA)*

During the 8 October 1953 public demonstration of the atomic cannon at Aberdeen Proving Ground, the T10 transporter is hauling the piece up a slope. When loaded with the atomic cannon, the Heavy Artillery Transporter T10 had the capability of driving up a grade of 30 percent, or a slope of 17 degrees. *(NARA)*

Chapter Four

Atomic Cannon in Service

Li'l Adam is in traveling position on its transporters at the East Firing Range, Fort Sill, on 1 April 1953. This is the view of the driver of the rear transporter, an M250 4 x 4 truck. The front transporter, the T10A, is coupled to the cannon, but only a small part of that vehicle, including the spare tire, is visible. Note the weld bead on the top center of the lifting fork of the T10A truck at the bottom of the photo. *(NARA)*

Privates Raymond Figures and Matthew Brown, of A Battery, 867th Field Artillery Battalion, pose while operating the left hydraulic leveling jack on an atomic cannon at Fort Sill on 19 March 1953. There were three jacks on the cannon, and they were used to adjust the lateral level of the top carriage of the gun with reference to the firing base (the turntable on which the front end of the top carriage pivoted). *(NARA)*

Corporal James Brannan, an artillery mechanic with A Battery, 867th Field Artillery Battalion, adjusts a cross-leveling jack on a 280mm Gun T131 at Fort Sill on 19 March 1953. Next to the head of his screwdriver is a plaque showing the direction for cranking the jack up or down. *(NARA)*

Another one of the 280mm atomic cannons at Fort Sill, nicknamed Big Ike, crosses a bridge en route to the East Firing Range on 1 April 1953. The front transporter truck, the T10A, is to the left; the USA number on its driver's door is 60146779. The 280mm cannon barrel is retracted, with the breech supported over the rear of the top carriage of the gun mount. *(NARA)*

Big Ike negotiates a difficult turn during an exercise at Fort Sill in April 1953. For all of its bulk and length, the atomic cannon and transporters package was quite maneuverable, thanks to the independently steered front and rear trucks. On the front bumper are markings for the 26th vehicle, Battery B, 264th Field Artillery Battalion. *(NARA)*

U.S.ARMY
ORDNANCE

(**Opposite, above**) Crewmen crouch in the foreground as a 280mm Gun T131 fires a practice round. To their front is a generator trailer that supplies power to systems on the cannon. Behind the embankment to the right of the atomic cannon is a 240mm Howitzer M1. The 280mm Gun T131 would not be standardized as the 280mm Gun M65 until around 1960. *(NARA)*

(**Opposite, below**) Members of the press are receiving a briefing on an atomic cannon at the Mainz Ordnance Depot in West Germany on 21 October 1953. The T10B rear transporter truck bears US Army registration number 60147660, while the T10A truck to the front of the cannon is USA number 60157659. *(NARA)*

(**Above**) On one of the 280mm Guns T131 shown to the press at Mainz on 21 October 1953, details of the equilibrator mechanisms in the front end of the top carriage are visible, along with other features such as the gun barrel, cradle, and recoil/recuperation cylinders on the bottom of the cradle. *(NARA)*

(**Opposite, above**) An atomic cannon crew of A Battery, 265th Field Artillery Battalion, are engaged in lowering the cannon and carriage from the T10 transporter near Baumholder, Western Area Command, West Germany, on 9 February 1954. The view is from the side of the T10A at the front. This was part of a simulated firing drill during a field problem. *(NARA)*

(**Opposite, below**) During the field problem near Baumholder on 9 February 1954, an atomic cannon has been emplaced and the barrel elevated. The atomic cannon crews were trained to defend their piece if necessary; note the stacked rifles to the right. To the left of the gun breech a crewman is standing next to the panoramic sight. *(NARA)*

(**Above**) The driver's door of a T10A transporter truck is open as the crew prepares to move an atomic cannon. The crew and vehicle were assigned to C Battery, 265th Field Artillery Battalion, and the scene was near Baumholder on 10 February 1954. The driver is wearing a headset and microphone in order to communicate with the driver of the T10A at the rear of the atomic cannon. Note the clearance flag on the side of the bumper. *(NARA)*

An atomic cannon of A Battery, 868th Field Artillery Battalion, is being transported through the gate of Patch Barracks at Seventh Army Headquarters, Vaihingen, West Germany, on 24 February 1954. The T10A 4 × 4 heavy-gun-lifting truck to the left is marked with USA number 60147659 and bears license plate number T-79487. *(NARA)*

An atomic cannon assigned to A Battery, 868th Field Artillery Battalion, is demonstrated to troops at Patch Barracks on 24 February 1954. The T10B rear transporter truck has been detached from the carriage, and the cannon barrel is being hauled into firing position. This is being accomplished by a cable from the winch, routed around the sheave atop the gun cradle, between the trunnions and back to the top of the gun breech. Stenciled artwork is on the center part of the top carriage. *(NARA)*

The atomic cannon has been emplaced on the pavement and its barrel set at high elevation, at Patch Barracks on 24 February 1954. To the left of the piece is an ammunition cart with a projectile on it. To the left of the cannon in the background is the T10B rear transporter truck for the piece. *(NARA)*

NIGHTMARE ALICE

(**Opposite, above**) The crew of an atomic cannon at Patch Barracks demonstrate the hoisting of a projectile on the same day the preceding photos were taken. The two men at the center are on the hand crank for the loading davit, while crewmen to the left guide the projectile. The projectile is clamped in a set of shot tongs attached to the davit cable. *(NARA)*

(**Opposite, below**) Private 1st Class Donald A. Devers of Pittsburgh, Pennsylvania, a member of A Battery, 868th Field Artillery Battalion, based in the Western Area Command of West Germany, adjusts the left cross-leveling jack on a 280mm Gun T131 nicknamed Nightmare Alice, on 14 April 1954. Towards the lower right is the crank fixture for the center cross-leveling jack. *(NARA)*

(**Above**) Angel is seen close-up with crewmen posing by it during Operation Flashburn at Fort Bragg on 26 April 1954. Artwork is to the rear of the 'Angel' inscription, showing an angel next to a devil's head over a nuclear mushroom cloud. A tarpaulin is draped over the forward end of the 280mm cannon barrel. *(NARA)*

Two crewmen remove a lifting ring from the point of a 280mm projectile on an ammunition cart near an atomic cannon. Note the Maltese-cross-type design on the manhole cover on the side of the top carriage of the cannon, and the 'U.S.A.' painted on the gun cradle. *(NARA)*

(**Opposite, above**) At the firing range at Baumholder, crewmen of the 868th Field Artillery Battalion prepare to hoist a 280mm projectile to an atomic cannon. From left to right they are: Corporal William F. Gulley, Sergeant Donald J. Miller, Private 1st Class Bivona Biagio, and Corporal Carwin Verbeek. The photo was taken on 9 April 1954. Shot tongs at the end of the winch cable are fastened to the projectile, and Private 1st Class Biagio is grasping the winch hand crank. *(NARA)*

(**Opposite, below**) It took two men to operate the winch crank for the loading davit, and two members of the crew seen in the preceding photo, Private 1st Class Biagio and Sergeant Miller, are working the crank handle to lift the projectile while Corporal Verbeek steadies the tip of the projectile. The boom of the loading davit is supported at the operating angle, shown here, by two steel-cable stays attached to the davit staff and the outer end of the boom. *(NARA)*

(**Opposite, above**) On the firing platform of the same atomic cannon, a projectile is on the rammer trough, prior to being rammed into the breech. In the foreground, Corporal William F. Gulley holds the removable lifting ring for the 280mm projectile while Private 1st Class Joseph D. Daley is about to place a firing pin into the projectile. In the left background, Sergeant Donald J. Miller stands next to the panoramic telescope, while on the other side of the gun breech is Corporal Guy L. Wilburn. *(NARA)*

(**Opposite, below**) Private 1st Class Bivona Biagio, 868th Field Artillery Battalion, rams ammunition into the breech of an atomic cannon the hard way, using a hand crank to manually operate the rammer mechanism, which usually was powered hydraulically. On the rammer trough are several bags of propellant. Hooked to the firing base under the forward end of the top carriage of the cannon is what looks like a pole; this was part of a stabilizing anchor for the firing base. *(NARA)*

(**Above**) On an atomic cannon assigned to the same unit as in the preceding photos, a projectile and powder bags have been rammed into the breech, and two crewmen, Private 1st Class John W. Allen and Private 1st Class George P. Harmotto, are lifting the breechblock to close it. To the right is Sergeant William E. Tucker. The interrupted-screw design of the breechblock is visible. *(NARA)*

(**Opposite, above**) A 280mm projectile is on the rammer trough of an atomic cannon at a firing range at Baumholder on 9 April 1954. One of the crewmen has just removed the firing pin from the projectile. The cable on the ground to the left of the cannon provides electrical power to the piece from a generator trailer. *(NARA)*

(**Opposite, below**) Corporal Guy L. Wilburn of Ava, Illinois, assistant gunner of an atomic cannon of the 868th Field Artillery Battalion, connects a firing lanyard to a triggering mechanism towards the right rear of the top carriage during an exercise at the firing range at Baumholder on 9 April 1954. The design of the firing mechanisms on this part of the gun carriage would change over time. *(NARA)*

(**Above**) Two corporals of A Battery, 868th Field Artillery Battalion, pull a winch cable on a T10A front transporter truck for an atomic cannon in the Western Area Command of West Germany on 14 April 1954. Note the spotlight on a flexible mount on the side of the winch assembly. Towards the bottom of the side of the winch is the hydraulic motor that drove the winch. *(NARA)*

(**Opposite, above**) Atomic cannons of A Battery, 867th Field Artillery Battalion, are emplaced and their crews are preparing them for action during an exercise in the Western Area Command, West Germany, on 28 April 1954. Two of the 280mm cannons are readily visible, and the forward end of top carriage of a third one is in the background, below the gun barrel of the second piece. *(NARA)*

(**Opposite, below**) The Radar Section of A Battery, 868th Field Artillery Battalion, was tasked with tracking the trajectories of projectiles fired by the battalion's atomic cannons while attached to the Western Area Command in West Germany. As seen in a 14 April 1954 photograph, Private 1st Class Fale R. Morrison and radar operators Corporal Jimmy T. Young and Corporal Donald W. Ogborn pose next to a radar indicator and control panel. In the background is a radar antenna on a trailer. *(NARA)*

(**Above**) The artilleryman at the center facing the photographer has just pulled the lanyard to fire a round from an atomic cannon of A Battery, 867th Field Artillery Battalion, at a firing range in West Germany on 28 April 1954. Stuck in the ground along the rear of the top carriage is a red and white aiming stick used to align the cannon for purposes of indirect fire. *(NARA)*

This still photograph was taken on 21 May 1954 while a Signal Corps crew was making a 'service of the piece' training film on the 280mm Gun T131. The cameramen on the platform to the left of the cannon are shooting footage as members of A Battery, 216th Field Artillery Battalion, serve the piece. The two men on the T10A truck in the foreground are preparing a winch cable running to the rear of the cannon breech, in order to retract the cannon barrel to the travel position. *(NARA)*

In late August 1954 this atomic cannon and T10 transporter skidded off a road and into a ditch at Murnau, Germany, causing severe damage to the rear transporter (foreground), and, presumably, the cannon and the front transporter. Army engineers brought in recovery equipment and worked around the clock to bring the 85-ton cannon into an upright position so it could be hauled away.

(**Above**) Civilians and a horse team, far right, have been halted as a 280mm Gun T131 assigned to A Battery, 264th Field Artillery Battalion, is being transported off a main route to a firing position during US Army exercises in West Germany on 25 October 1954. The T10A/XM249 truck to the right is USA number 60147641. A canvas cover is lashed over the cannon barrel from the front of the breech to the gun cradle, to protect the unpainted steel from the elements. *(NARA)*

(**Opposite, above**) This 280mm atomic cannon is on display at the Anniston Ordnance Depot in Alabama in January 1955 as part of a US Army effort to publicize the movie *This is your Army*, which was being shown in theaters in the city of Anniston. The front transporter vehicle, by now designated the Truck, Gun-lifting, Heavy, 4 × 4, Front M249, bears US Army registration number 60146855, while the M250 to the rear is USA number 60146756. Typically, M249 and M250s bore consecutive USA numbers. *(NARA)*

(**Opposite, below**) Members of the 867th Field Artillery Battalion in West Germany are pulling maintenance on a 280mm atomic cannon nicknamed Archie, on 20 January 1955. The men to the rear of the cannon carriage are leaning into a rammer staff, to clean the bore of the big gun. Note the GI sitting on the muzzle of the piece. *(NARA)*

USA
USA

507222

(**Opposite, above**) An atomic cannon assigned to A Battery, 692nd Field Artillery Battalion, is being transported at Fort Sill on 15 June 1955. The M250 rear truck to the left bears USA number 60146770. Prominent on top of the engine compartment are the early-type fan shroud extension and exposed muffler. 'U.S. ARMY' is painted on the gun cradle and the top carriage. *(NARA)*

(**Opposite, below**) In the first of a sequence of photos showing preparations to move an emplaced 280mm Gun T131 'atomic cannon' at a training area near Grafenwöhr, in the Southern Area Command, West Germany, on 8 August 1955, camouflage netting masks the position of the piece. Out in the open to the front of the cannon is the M250 rear transporter truck assigned to the cannon. *(NARA)*

(**Above**) Artillerymen remove camouflage netting from the emplaced atomic cannon, which was assigned to B Battery, 216th Field Artillery Battalion, in preparation for moving the piece to another firing position. The netting constituted several nets, supported by poles and the gun barrel. The second man from the left is bending over a stack of dome-shaped wire cages and poles. These cages were attached to the poles that supported the netting, the cages forming a surface to hold the nets. *(NARA)*

(**Opposite, above**) The camouflage netting has been removed from the atomic cannon. A canvas cover is fitted over the panoramic telescope mounted on the side of the left trunnion, and a cover is also present over the muzzle of the cannon. Army doctrine called for atomic cannon crews to 'shoot and scoot': that is, to be prepared to emplace and fire the pieces and quickly move to another firing position before the enemy determined the cannon's location. *(NARA)*

(**Opposite, below**) The first step in putting the cannon into travel configuration was to back the front truck, the M249, up to the rear of the carriage so that the two prongs of the lifting fork at the rear of the truck could be positioned into sockets on top of the rear of the top carriage of the cannon. The closest artilleryman on the ground is signaling the driver as he backs the truck up. *(NARA)*

(**Above**) The front tips of the lifting fork of the M249 truck have been inserted into the sockets on the top carriage, and the artilleryman on the ground is connecting the bottom of one of the hoists to a hook at the lower rear of the carriage. The idea was to make a stable connection between the M249 and the cannon carriage before the M249 winched the gun barrel, to the rear, to its travel position. *(NARA)*

(**Above**) This photo was taken from the rear deck of the M249 front transporter truck that will haul the rear end of the cannon carriage. Attached to the retraction eyebolt on the upper rear of the breech ring of the 280mm cannon is the winch cable, which soon will retract the barrel to its travel position. First it was necessary to disconnect the barrel from the recoil/recuperation cylinders. *(NARA)*

(**Opposite, above**) The winch cable, fastened to the eyebolt on the rear of the breech ring, is viewed from a different angle. The loop at the end of the winch cable is attached to the eyebolt by means of a shackle. The stencils on the side of the breech and on the top carriage to the rear of the breech give the costs of the respective components, to impress on the crew the necessity of taking good care of the piece. *(NARA)*

(**Opposite, below**) Crewmen pose for the camera as the winch on the M249 is drawing the 280mm cannon barrel to the rear. To the left, on the side of the front bumper, is one of two hooks used for storing the bottom of the hoist pistons when they were not being used. Visible on the side of the winch are the flexible spotlight, the winch transmission shift lever, and towards the bottom, the hydraulic motor and the four-way valves. *(NARA)*

(**Above**) The 280mm cannon barrel has been winched back to its retracted position for travel. The platforms along the top carriage of the cannon have been folded and secured. To the far right, an emergency-brake jumper line has been attached to couplings on the lifting fork and the carriage. *(NARA)*

(**Opposite, above**) The cannon barrel has been retracted and secured and the gun and carriage hoisted and secured to the M249 and M250 trucks. A tarpaulin has been rigged over the rear of the carriage and the unit is ready for transport during a training exercise at Grafenwöhr on 8 August 1955. The USA number on the M249's side door and unit markings on the bumper have been obscured. *(NARA)*

(**Opposite, below**) An M249 truck, USA number 60147661, is being maneuvered near the rear end of the carriage of an atomic cannon assigned to the 59th Field Artillery Battalion. In the background are two more trucks from a T10 transporter; an M249 to the left and an M250 rear truck visible below the lifting fork of the M249 in the foreground. Note the warning stripes on the front tips of the loading forks. The scene was the training area at Grafenwöhr on 18 August 1955. *(NARA)*

U.S.A.
60147661

US
ARM

(**Opposite, above**) Members of A Battery, 59th Field Artillery Battalion are cleaning and lubricating a 280mm atomic cannon prior to testing the piece at the Grafenwöhr training area, West Germany, on 18 August 1955. Two of the men are brushing and rubbing oil on the unpainted part of the barrel, which had to slide freely in the enclosing cradle during recoil and recuperation. To the left, the breechblock is open, allowing a good view of its interrupted-screw design. *(NARA)*

(**Opposite, below**) This photo shows the same cannon and crew as the preceding photo, although from the left rear of the piece. The two lower platforms and their steps have been lowered, but the center, higher platform is in the travel position. A cover is over the panoramic telescope on the trunnion. *(NARA)*

(**Above**) A member of A Battery, 59th Field Artillery Battalion, is cleaning and oiling the breech of an atomic cannon at Grafenwöhr training area on 18 August 1955. Below the left side of the breech is the left retraction wheel and gun-retraction stop. The two retraction wheels allowed the breech-end of the gun to be retracted for travel (as seen here) and pulled forwards on two retraction rails. The breech was pinned to the retraction stops to immobilize it during travel. *(NARA)*

USARMY
U.S.
ARMY

(**Opposite, above**) An atomic cannon is being maneuvered between the two trucks of a T10 transporter. The piece probably is being lowered to the ground, as there are no visible impressions in the soil below the firing base (foreground) and float (background) of the cannon's bottom carriage. On the cowl of the M250 rear truck to the left is the bridge-classification number 35. *(NARA)*

(**Opposite, below**) Crates containing six 280mm cannon tubes are stowed in the hold of the cargo ship USNS *Marine Fiddler* (T-AK-267) at Naha, Okinawa, on 1 August 1955. These tubes had been removed from atomic cannons of the 663rd Artillery Battalion (280mm) on Okinawa and stowed separately. The muzzle ends of the tubes are protruding from the far ends of the crates. For a sense of scale, note the two trucks stored to the far right. *(NARA)*

(**Above**) On the same date the preceding photo was taken, a 280mm Gun T131 'atomic cannon' on a shipping pallet is being unloaded from USNS *Marine Fiddler* at Naha, Okinawa. The cannon barrels would be installed later. Like the 280mm gun tubes in the preceding view, this cannon was destined for the 663rd Artillery Battalion (280mm). It required a gigantic 150-ton crane to hoist the atomic cannon from the ship. Shipping information is stenciled on the side of the top carriage. *(NARA)*

(**Above**) An M249 4 x 4 heavy gun-lifting truck, US Army registration number 60146761, is being unloaded from USNS *Marine Fiddler* at Naha on 1 August 1955. Note the light-colored protective material applied over the grilles on the side of the engine compartment. The spare tire and the spare-tire davit are in place to the rear of the cab. *(NARA)*

(**Opposite, above**) Once this M249 truck and atomic cannon, minus the gun barrel, were deposited on the dock next to USNS *Marine Fiddler* at Naha, efforts were made to couple the M249 to the cannon carriage. The tips of the lifting fork have been maneuvered into the sockets on the top of the cannon carriage. *(NARA)*

(**Opposite, below**) The M250 rear truck of the T10 transporter as well as the M249 front truck have lifted the atomic cannon carriage and cradle into place, and the entire load has been secured for travel. The M250 is marked with USA number 60146766. Note the protective material taped over the round opening on the end of the gun cradle. The atomic cannons that USNS *Marine Fiddler* delivered to the 663rd Artillery Battalion (280mm) at Naha, Okinawa, on 1 August 1955 were the first such weapons deployed to the Far East. *(NARA)*

Members of the 663rd Artillery Battalion (280mm) are mounting the barrel onto an atomic cannon at the Nashville Motor Pool on Okinawa on 11 August 1955. Two crawler cranes are necessary to lift the heavy barrel in place. It will be necessary to insert the muzzle end into the opening in the cradle and push the barrel forward. *(NARA)*

Two truck cranes are maneuvering a 280mm gun tube that is to be mounted on the atomic cannon carriage suspended between the T10 transporter to the left. Pieces of wooden boards are arranged around the barrel where the two slings are positioned to protect the sensitive surface from damage. *(NARA)*

An atomic cannon assigned to B Battery, 613th Field Artillery Battalion, rolls through the gate of Fort Polk, Louisiana, on 27 October 1955. This piece and its crew would soon be participating in Exercise Sagebrush, which would involve the simulated firing of conventional and atomic rounds. The M249 truck in the foreground is registration number 60146785. *(NARA)*

A T10 transporter hauling an atomic cannon is parked in a woodland. The registration number on the side door of the M249 truck is indistinct, as are the unit markings on the bumper. A circle with the number 98 on it is below the right headlight array, and the number 54 is in a circle on the door. The small insignia on the front of the cab between the warning light and the left headlight array is a two-colored rectangle with an X across it. *(NARA)*

An artilleryman is inserting a lock wedge in the aperture through the locking-wedge block on the top carriage of the cannon and the similar locking block on the underside of the lifting fork, called a 'keeper' in the tech manuals. The locking wedges inserted in the blocks and keepers at the four corners of the carriage secured the cannon in place for transport. *(NARA)*

The hydraulic hoists on the lifting fork of an M249, recognizable by the shapes of the fork and the front bumper, are shackled to hooks on the top carriage of an atomic cannon. This photo gives a particularly good idea of the shapes and the relative positions of the locking-wedge blocks on the upper corners of the top carriage and, above them, the keepers on the undersides of the lifting forks. Alongside the locking-wedge block are stored a locking wedge, wedge-fastening key, and key-locking pin. *(NARA)*

U.S.A.

(**Opposite, above**) An atomic cannon is suspended between the two trucks of a T10 transporter, with the front of the gun and the carriage to the right. The hoists are still shackled to the top carriage of the cannon, and the air hoses are disconnected from the carriage. *(NARA)*

(**Opposite, below**) A Kenworth M250, US Army registration number 60146808, is seen from the left side at a firing range. In the right background, partially hidden by the M249, is an emplaced atomic cannon. To the rear of the cab is the engine compartment, housing the Continental AO-895-4 four-cylinder, air-cooled gasoline engine. *(NARA)*

(**Above**) In a view from the rear of M249 registration number 60146807, which likely was the mate of the M250 in the preceding photograph, the design of the cross member of the lifting fork, with its ten lightening holes, is seen to good advantage. Atop the lifting fork to the sides of the cross member are the so-called lifting brackets at the tops of the lifting cylinders. *(NARA)*

(**Above**) Two artillerymen next to an atomic cannon are grasping the handle of an ammunition cart with a 280mm projectile on it. These carts were standard-issue accessories for an atomic cannon and were used to haul projectiles to the loading davit mounted on the left side of the gun carriage. *(NARA)*

(**Opposite, above**) Crewmen of an atomic cannon demonstrate using the loading davit to lift a 280mm projectile up to the rammer assembly for loading into the gun. Normally it took two men to operate the davit winch, but here one man is doing it. Clutching the projectile is a set of shot tongs, which uses the weight of the projectile to clamp itself tightly around it. The man on the platform has his left hand on the staff handle to steer the davit. *(NARA)*

(**Opposite, below**) An army cameraman on the rear of the top carriage of an atomic cannon is filming an ammunition-loading exercise, most likely for use in a training film or documentary production. The rear of the carriage and the float are mud-spattered: not exactly conducive to traversing the piece, which depended on clean slides on the float. *(NARA)*

This is a trailer-mounted portable generator set for providing electrical power to an atomic cannon. The doors to the control panel are open, showing the various placards, switches, dials, and gauges inside. At the bottom are various receptacles with covers. To the lower right is the power cable, which is routed to the receptacle box on the atomic cannon. *(NARA)*

On the right side of an atomic cannon carriage, an artilleryman has his hand on the elevating speed-control handle. The box to the left (i.e., rear) of that handle contains the equilibrator motor control switch, while the gauge to the right of the handle is for the equilibrator pressure. To the far right is part of the elevating hand wheel. *(NARA)*

On the left side of an atomic cannon of the 692nd Field Artillery Battalion at Fort Sill in June 1955, a gunner operates the traversing hand wheel while sighting through the M12A7C panoramic telescope. This type of telescope was a standard US Army type used on other artillery pieces: for example, the 155mm Howitzer M114. It was used to lay the cannon in azimuth and could be used for direct or indirect fire. The gunner is peering through the eyepiece. At the top of the telescope is a rotating head containing the objective. *(NARA)*

Crewmen of three atomic cannons of the 59th Field Artillery Battalion man their guns for a commander's inspection at Pirmasens, West Germany, on 21 April 1956. Comic-book characters are the subjects of nicknames and artwork on the front ends of the cannon cradles. From front to rear they are Sad Sack, Bugs, and Beetle Bailey. *(NARA)*

U.S.ARMY
COOL
CAT
RMY

(**Opposite, above**) More atomic cannons of the 59th Field Artillery Battalion are displayed at the battalion motor park at Pirmasens on 18 April 1956. The nicknames of these pieces are, front to rear, Cool Cat, unknown, and Able Annie. All three cannons are in training-fire position. Two shop trucks are parked between the cannons. *(NARA)*

(**Opposite, below**) Private George P. Burke touches up the feline-in-an-ice-cube artwork of Cool Cat, an atomic cannon assigned to the 59th Field Artillery Battalion at Pirmasens on 29 April 1956. All of this battalion's atomic cannons had nicknames. *(NARA)*

(**Above**) Crewmen of Sad Sack, a 280mm atomic cannon of the 59th Field Artillery Battalion, are in the process of hoisting a projectile up for loading into the gun during a practice drill at Pirmasens on 10 April 1956. Two men operate the davit-winch crank while another steadies the projectile. *(NARA)*

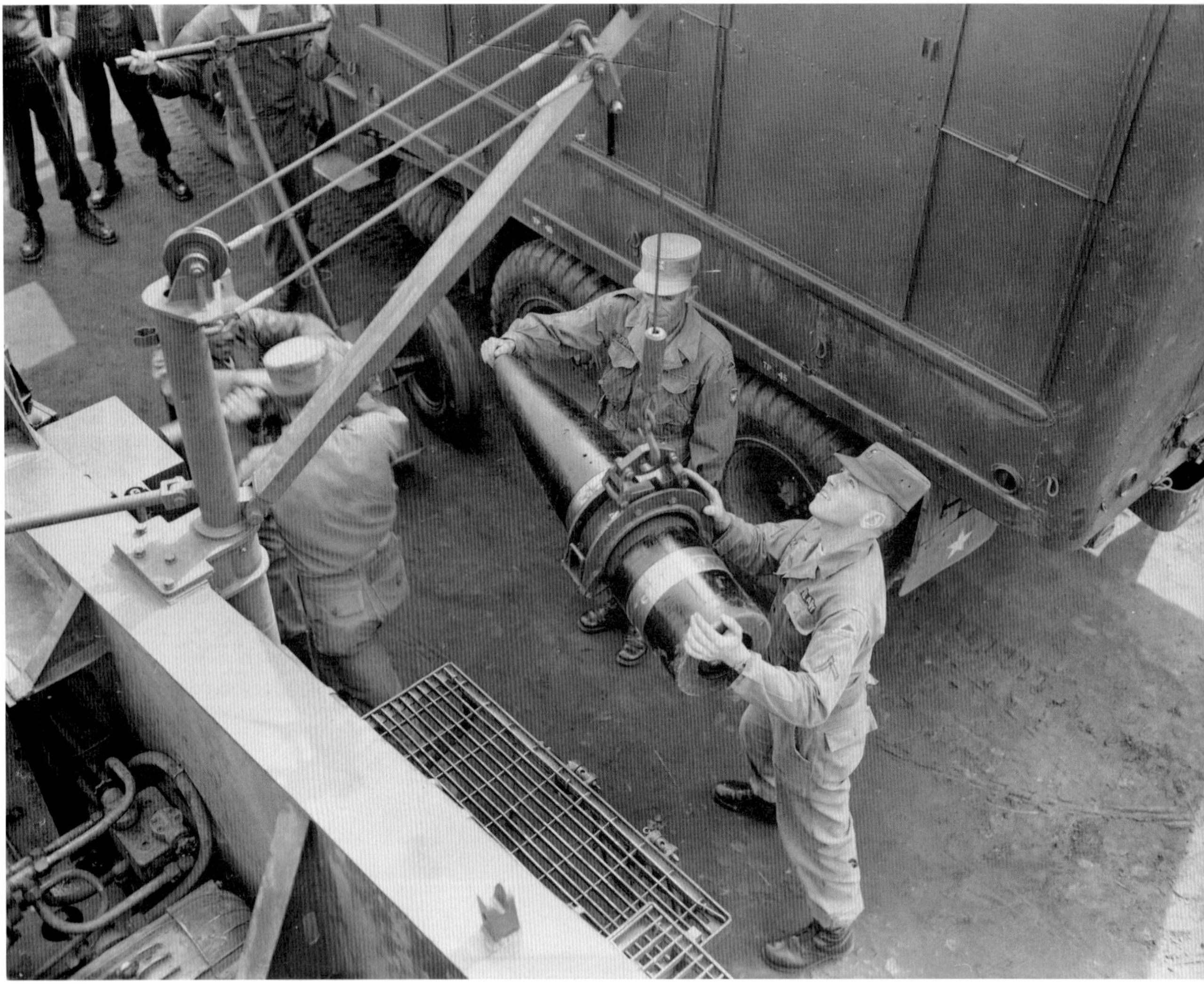

(**Above**) Standing between an atomic cannon of the 59th Field Artillery Battalion and a shop truck, two crewmen brace a 280mm projectile as two other men crank the davit winch. The view is from the left side of the gun breech. In the background a crewman is holding onto an ammunition cart. *(NARA)*

(**Opposite, above**) The crew of a 280mm atomic cannon of the 868th Field Artillery Battalion prepares to fire the piece during a training exercise at Grafenwöhr on 28 May 1956. The men are busy leveling the carriage, making electrical connections, and preparing the gun for action. *(NARA)*

(**Opposite, below**) In Germany, with its many large rivers, field artillery battalions armed with atomic cannons had to ensure that during wartime they would be able to transport their pieces across bodies of water even if bridges had been destroyed or were unsuitable. Thus these units occasionally practiced ferrying their cannons across rivers. Here a British raft is ferrying an atomic cannon of B Battery, 868th Field Artillery Battalion, across the Rhine during Exercise Kravel on 21 June 1956. *(NARA)*

(**Above**) The atomic cannon seen in the preceding photo is disembarking from the British raft at Krefeld, West Germany, on 21 June 1956. This was part of an experiment to determine if this type of raft could safely transport the atomic cannon. The placard tied to the bumper is in German and translates to, 'Danger 280mm Cannon/28 meters long'.

(**Opposite, above**) On 21 June 1956, a British landing ship, tank (LST), L407, also was used in an experiment to ferry an atomic cannon across the Rhine at Krefeld. The craft was manned by Royal Marines. *(NARA)*

(**Opposite, below**) L407 makes a hard turn to starboard while ferrying the atomic cannon across the Rhine at Krefeld on 21 June 1956. The T10 transporter and the cannon had been backed into the landing craft, positioned to be able to drive forward off the ramp upon reaching shore. *(NARA)*

L407

L407

US Army military equipment, including an atomic cannon, is on public display at an unidentified location, probably in Europe. To the rear of the cannon are the two trucks of the T10 transporter, with the rear end of the rear truck closest to the camera perspective. To the right of the closer transporter truck is the portable generator for the atomic cannon. *(NARA)*

An atomic cannon is temporarily situated on a dirt road under a natural camouflage cover of trees. During field exercises, these guns were moved around to simulate the fast-moving shoot-and-scoot operations that they would conduct in actual wartime, to keep a step ahead of enemy aircraft and artillery spotters. When the gun was parked between firing missions, it was essential to camouflage it, either with netting or under natural cover such as this. *(NARA)*

U.S.
66

(**Opposite, above**) Artillerymen of the 1st Section of B Battery, 59th Field Artillery Battalion, prepare to load a projectile into the atomic cannon nicknamed Bugs at a site near Vilseck, West Germany, in early May 1956. The two men to the right are holding a hand rammer, used to ram ammunition into the breech in the event of the failure of the power or manual rammer system. Note the number 66 near the rear of the top carriage. *(NARA)*

(**Opposite, below**) Eight men on the ground and two more on the loading platform of Bugs are leaning into the rammer staff during a hand-ramming exercise. It only took the hydraulically powered rammer a few seconds to shove a projectile into the breech; it took much longer with a rammer staff. *(NARA)*

(**Above**) A corporal on the loading platform of an atomic cannon of the 59th Field Artillery Battalion is lending his muscle to the hand-ramming of a 280mm projectile into an atomic cannon on 8 May 1956. The powder bags that went in after the projectile were much lighter and easier to load. *(NARA)*

U.S.
ARMY

(**Opposite, above**) Two crewmen on the right side of the carriage of Bugs, an atomic cannon assigned from the 59th Field Artillery Battalion, are operating the elevating hand wheel during an exercise near Vilseck on 8 May 1956. Other men are preparing the piece for firing. Note the aiming sticks placed in the ground near the front corners of the carriage. *(NARA)*

(**Opposite, below**) A 280mm atomic cannon of A Battery 804th Field Artillery Battalion is being prepared for a march-order drill exercise near Grafenwöhr on 7 August 1956. The M249 truck to the left is being backed up to the rear of the cannon carriage, following which the winch on that truck will retract the cannon barrel to the travel position. Next, the M250 truck to the right will move in, and both trucks will hoist the cannon to travel position. *(NARA)*

(**Above**) The crew of the same atomic cannon in the preceding photo takes a five-minute break during a road march on 7 August 1956. The number 31 is on the right side of the bumper of the M249, indicating that it was the 31st vehicle in the order of march of its battalion. Two 5-gallon liquid containers are on holders on the side of the engine compartment. *(NARA)*

(**Above**) An artilleryman of the 59th Field Artillery Battalion is adjusting the dust cover on the muzzle of a 280mm atomic cannon. This essential piece of equipment prevented foreign matter from entering the barrel and also helped protect the bare metal of the bore from the effects of moisture. Hitched to the rear of the 6 × 6 cargo truck to the right is a portable generator for the atomic cannon. *(NARA)*

(**Opposite, above**) An M249 truck, USA number 602146769, part of a T10 transporter for an atomic cannon, is loaded on a flatcar at Ruston, Louisiana, on 19 March 1957. This vehicle, attached to A Battery, 692nd Field Artillery Battalion out of Fort Sill, Oklahoma, was being shipped to Fort Polk, Louisiana, where it would participate in Exercise King Cole. *(NARA)*

(**Opposite, below**) Three complete atomic cannons and their T10 transporters, assigned to the 692nd Field Artillery Battalion, were loaded on railroad flatcars at Ruston when this photo was taken on 19 March 1957. To the far left is the lifting fork of one of the transporter trucks. On the next car is an atomic cannon with the barrel removed. In the background are more cannons and trucks. *(NARA)*

U.S. ARMY
60146769
U.S. ARMY
60146769
A30A
GREAT NORT HERN 65470
AB BRAKE
NO.15 BRAKE BEAMS
52 Ft

52 Ft

The train loaded with atomic cannons and transporters of the 692nd Field Artillery Battalion is seen from another angle at Ruston. At the bottom is a lifting fork of one of the transporter trucks. On the next car is an atomic cannon minus its barrel. Wooden beams were bolted to the floor of the flatcar to brace the cannon, and there were various hold-down rods and beams to immobilize the gun carriage. *(NARA)*

(**Opposite, above**) For Exercise King Cole in Louisiana in 1957, the 280mm gun barrels were removed from the atomic cannons and shipped on their own flatcars, as seen here. Spanning this photo are two flatcars, each with a cannon barrel secured to it. On the nearest car, the breech of the cannon is to the left. Above that barrel is the gun cradle of an atomic cannon on a flatcar on the adjacent track. To the far left is an M249 truck.

(**Opposite, below**) Five days after the preceding sequence of photos of atomic cannons on flatcars at Ruston was taken, one of the cannons has been completely assembled along the tracks and is ready for transport to Fort Polk for Exercise King Cole. The view is along the left side of the gun carriage looking to the rear, with an M250 truck hitched to the rear of the carriage. Another M250 and atomic cannon is next to this one. *(NARA)*

L & N
222 17

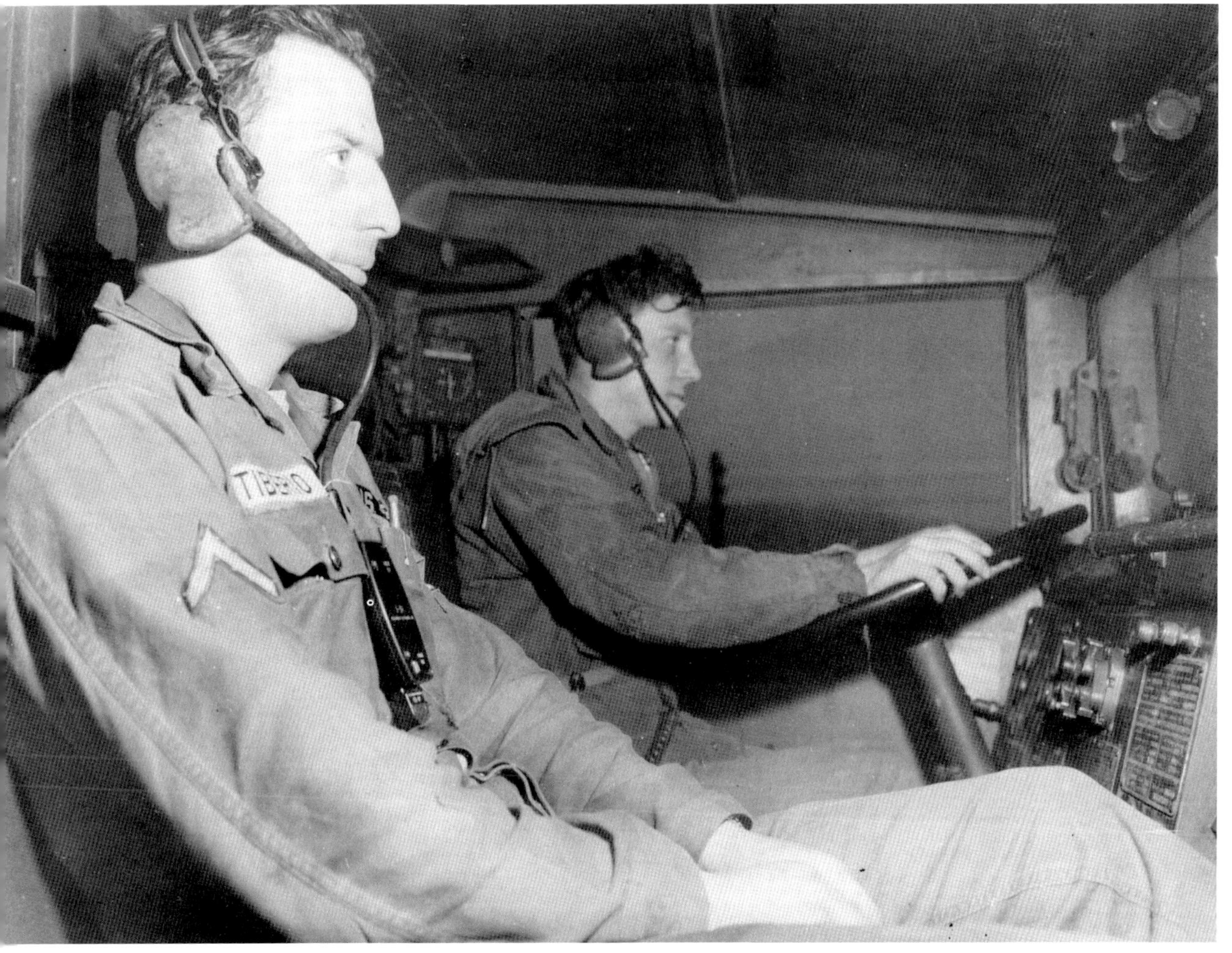

(**Opposite, above**) A 280mm atomic cannon of the 3rd Gun Battalion, 81st Artillery, 42nd Field Artillery Group, is emplaced in a firing position as part of a demonstration of US Army weapons arranged for officers of the German Army's I Corps at Gonsenheim-Mainz, West Germany, on 28 October 1958. To the right are two 6 x 6 cargo trucks for transporting the crew. Between the trucks is a portable generator set. To the left of the left truck is part of the lifting fork of a T10 transporter truck. *(NARA)*

(**Opposite, below**) An atomic cannon has just fired a 280mm projectile during a field exercise at Grafenwöhr on 17 May 1958. The gun and crew were part of the 265th Field Artillery Battalion, and it had taken them exactly thirty minutes from the time they arrived at this firing site until they fired the first shot. *(NARA)*

(**Above**) In an M250 rear transporter truck of an atomic cannon, the assistant driver, Private 1st Class Charles L. Tigerio, left, receives information on upcoming road turns from the crew of the M249 at the front of the cannon. Meanwhile the driver, Private Roy L. McMurthy, receives information on movement and position from the driver of the M249. These men were with the 868th Field Artillery Battalion at Baumholder on 17 May 1958. *(NARA)*

U.S ARMY
60146795
98
7A-39F
3C-23

(**Opposite, above**) A 280mm atomic cannon and T10 transporter assigned to the 39th Field Artillery Battalion is parked adjacent to a wrecked building during an exercise in the Grafenwöhr Training Area on 12 October 1958. Note the large white recognition stars on the lifting forks of the transporter trucks in addition to the ones on the cab doors. On the bumper of the M249 is the number of the vehicle's order of march: 24.

(**Opposite, below**) Another atomic cannon and T10 transporter of the 39th Field Artillery Battalion is temporarily parked during exercises at Grafenwöhr Training Area on 12 October 1958. The M249 truck in the foreground is marked with US Army registration number 60146795. *(NARA)*

(**Above**) In mid-May 1958 an atomic cannon of the 265th Field Artillery Battalion fires a 280mm projectile during a field exercise near Grafenwöhr. The crewmen have vacated the cannon, taking cover nearby to avoid injury from the blast. *(NARA)*

IV
US ARMY 60146800

U.S. ARMY

(**Opposite, above**) This 280mm atomic cannon of the 38th Field Artillery is about to go into the Mainz Ordnance Plant for repairs in August 1959. Several mechanics are working on the bottom carriage of the gun. The M250 to the right is registration number 60146800. Next to the cannon is the Artillery Rebuild Branch building. *(NARA)*

(**Opposite, below**) Two members of A Battery, 3rd Gun Battalion, 82nd Artillery, Private 1st Class Raymond B. Martin, right, and Private Jack N. Turkman, prepare an atomic cannon for travel at a hiding area at Grafenwöhr on 15 August 1959. They are about to insert a wedge to lock the right front corner of the upper carriage of the cannon to the lifting fork of the M250 truck. Note the colored caution stripes on the tip of the lifting fork and the camouflage net over the barrel. *(NARA)*

(**Above**) Military dependents wave as an atomic cannon of B Battery, 3rd Gun Battalion, 84th Artillery returns from field maneuvers to Baumholder on 26 September 1959. This battery had the highest score during firing exercises, and the inscription on the front of the cab of the M249 truck proclaims this fact. *(NARA)*

(**Above, left**) Inside an ammunition truck, Sergeant 1st Class Edward Turner, battalion supply sergeant of the 3rd Gun Battalion, 82nd Artillery, inspects 280m projectiles destined for an atomic cannon conducting firing exercises at Grafenwöhr on 15 August 1959. At the top is a chain hoist on a rail for hoisting and moving the ammunition. *(NARA)*

(**Above, right**) As seen from the top carriage of an atomic cannon, the cable from the loading davit has been clamped to a 280mm projectile resting on an ammunition cart, and the two men at the bottom are operating the davit winch to hoist the projectile up to the rammer cannon. The projectiles weighed approximately 600 pounds. The photo was taken in the vicinity of Grafenwöhr on 23 July 1959. *(NARA)*

(**Opposite, above**) A group of GIs survey an M250 truck that has rolled over into a ditch along the Autobahn in West Germany before attempting a recovery. The rear of the truck is to the left, with the three air reservoirs visible on the underside. Newspaper articles as early as the fall of 1955 referred to the frequency of accidents in which atomic cannons and their transports rolled over or became mired in ditches during movements in Europe. Sometimes it took several days to recover the cannons and transports. *(William Eisenhauer via US Army Field Artillery Museum)*

(**Opposite, below**) This M250, identified in an original caption as belonging to C Battery (presumably of the 3rd Battalion, 82nd Artillery) photographed in West Germany in 1962, has a demolished cab top and is evidently the same vehicle as shown in the preceding photograph, after being recovered. Note the extensive white stenciling on the fuel tank between the wheels. *(William Eisenhauer via US Army Field Artillery Museum)*

35

Following a roll-over of an atomic cannon of the 38th Artillery in West Germany, the recovered wreckage was assembled. The large disc at the center is the firing base of the cannon. Behind it is an M249 truck, and to the right is the lifting fork and front left tire of an M250 truck. To the far left are the remains of a cab top and a muffler shield.

Personnel from the US Army Ordnance Corps presented a demonstration of this 280mm atomic cannon at Bolling Air Force Base, Washington DC, on 18 May 1956 as part of an Armed Forces Day program. Here, the rear tractor has been detached from the piece, but the front tractor is still hitched to it. *(NARA)*

(**Above**) This photo was taken a split second after the 280mm atomic cannon fired a powder charge during a simulated live-firing demonstration at Bolling Air Force Base in May 1956. The two crewmen to the right have retired a respectful distance from the cannon during firing. (**Below**) Moments later, a charge was detonated down range, simulating the impact of a 280mm shell fired by the atomic cannon. *(NARA)*

A group of T10 transporter trucks for atomic cannons are lined up at a base in West Germany. From left to right they are an M250, two M249s, and another M250. All of the M250s are parked with their rears to the camera. *(William Eisenhauer via US Army Field Artillery Museum)*

In West Germany in the winter of 1962–1963, an M250 truck with an atomic cannon coupled to it is parked at right angles to the cannon. The truck and cannon were assigned to C Battery, 82nd Artillery, and the truck was registration number 60147644. *(William Eisenhauer via US Army Field Artillery Museum)*

(**Above**) A 280mm atomic cannon is emplaced in a firing position in West Germany. Of interest are: the yellow and black caution stripes on the cover of the loading davit winch; the white cover on the AC magneto starter box to the rear of the davit winch cover; and the yellow and black caution stripes on the edge of the float, under the rear of the top carriage. *(William Eisenhauer via US Army Field Artillery Museum)*

(**Opposite, above**) The only time an atomic cannon ever fired a nuclear projectile was on 25 May 1953 during the test code-named 'Grable' in Operation Upshot-Knothole, a series of nuclear tests at the Nevada Test Site. Here, at its firing position at the Frenchman Flat test range at 8.30 am, the cannon has just fired a W9 nuclear projectile with a yield of 15 kilotons: the same yield as the Little Boy nuclear bomb that destroyed Hiroshima on 6 August 1945. *(NARA)*

(**Opposite, below**) The W9 nuclear projectile has detonated and the fireball is ascending. The projectile exploded 524 feet above the ground and 200 feet from its intended target, at a range of 6.25 miles from the atomic cannon. *(NARA)*

Personnel connected with the Grable test in Operation Upshot-Knothole, including the atomic cannon crew, took cover before the explosion. Here several soldiers have come out of cover to observe the mushroom cloud from the W9 nuclear projectile blast. *(NARA)*

The crew of the atomic cannon that participated in the Grable test on 25 May 1953 was based at Fort Sill and they are seen here posing next to their piece.
(US Army Field Artillery Museum)

Chapter Five

Atomic Cannon in Detail

Nicknamed Atomic Annie, the atomic cannon used in the Grable test of Operation Upshot-Knothole currently is preserved at Fort Sill along with a T10 transporter consisting of M249 and M250 trucks.

A surviving example of the 280mm Gun M65, or atomic cannon, is this example at the US Army Artillery Museum, Fort Sill. It is displayed with M249 (front) and M250 (rear) heavy-artillery transporters. In the foreground is the front truck of the two-truck transporter, originally designated the T10A, and standardized as Truck, Gun-lifting, Heavy, 4 × 4, Front M249. The truck, USA number 60146753, is in markings for Battery A, 867th Field Artillery Battalion.

On the right side of the cab, grab-handles are on the door and on the body to the rear of the door. Blackout and service clearance lights are on the front corner of the body. Below the running board is a boarding ladder. The engine compartment is to the rear of the cab, with two screen grilles on the side, to the rear of which is the winch. Note the dual rear wheels.

On the front of the M249 cab, a marker light and a service headlight are in a recess on each side. Between them is a recess that held a warning light. On the bumper, looking from left to right, are the cable guide for the winch cable when it was deployed to the front of the vehicle; the right tow eye and clevis, the front tow pintle (at the center of the bumper); and the left tow eye and clevis.

Details of the left side of the lifting fork of the M249 are available. Note how the lifting fork rests on the top of the rear bumper when in the transport position, with no cannon hoisted under it. The two lines and couplings on the side of the fork are for the brake and throttle connections from the driver's controls in the M249 to the M250 rear truck, since the driver of the M249 controlled the brakes and the air-operated throttle of the M250.

More details of the three air reservoirs are shown, with the underside of the front bumper located at the top of the photograph. Air lines are fitted to the fronts of the reservoirs. Two lateral brackets support the reservoirs, which are nestled between the right and left members of the chassis frame.

In another view of the underside of the M249 chassis looking aft, to the right are the left boarding ladder and, to the front of the tire, the steering gear. Towards the top right is a driveshaft between a gearbox at the bottom of the steering column (out of view) and the steering gear. Protruding through a square opening on the bottom of the frame adjacent to the steering gear is an arm connected to the drag link.

On the right side of the cab of the M249, to the front is a rear-view mirror bracket, with two marker lamps below it. The hinge on the front edge of the door is a piano-type hinge. Note the diagonal brace on the bottom rung of the boarding ladder, and the hex screw on the upright just below the upper rung.

The right side of the engine compartment is viewed close-up. The grilles are fabricated of woven-wire screen material. Retainers for the grilles are inside the compartment and are secured by the hex screws seen here.

The M249 4 × 4 heavy gun-lifting truck was equipped with a Tulsa 90 hydraulically-powered winch. It had a capacity of 60,000 pounds and was provided with 300 feet of 7/8-inch cable. The winch is seen from the right side. To the left (i.e., rear) of the top of the long lever (the winch transmission shifter) is the spotlight bracket, while to the right is the hoist-winch selector valve, below which is the bleed valve. To the left and right of the bottom of the winch transmission shifter is the winch hydraulic motor.

Below the winch hydraulic motor (top of the photo) are the front four-way valve (to the right) and the rear four-way valve (left). Below those valves are lines from the main hydraulic pump on the opposite side of the chassis and winch hydraulic motor lines.

In a view from next to the right side of the engine compartment, we see the winch drum, loaded with 7/8-inch cable, and the main hydraulic pump shift lever, in addition to the winch transmission shifter to the left. The box-shaped object to the far side of the winch drum is the hydraulic oil reservoir.

In a view taken from above the winch drum, facing left, to the right is the perforated shield for the muffler. To the upper left is the hydraulic oil reservoir, which served both the winch and the hydraulic hoist cylinders on the lifting fork. Nestled between the muffler and the reservoir is the hydraulic oil pump for the winch.

In a view from the rear of the spare-tire carrier (not visible in this photo), in the foreground, is the hydraulic oil reservoir, with a filler cap on top. Next is the winch drum, with winch cable and loop. On the far side is the winch motor. To the upper left is the right rear of the engine compartment and part of the muffler shield.

This perspective is from the left rear of the engine compartment, with part of the spare-tire carrier to the extreme lower right, the lifting fork to the right, the hydraulic oil reservoir to the center, and the winch to the left. Note the hydraulic lines routed on top of the lifting fork; these are high-pressure and low-pressure lines destined for the hydraulic hoist cylinders.

(**Above**) The Tulsa winch on the M249 4 x 4 heavy gun-lifting truck is seen from the rear, with the right side of the engine compartment and the cab in the right background, the fifth wheel for the lifting fork to the left, and the spare-tire davit in the background to the left of the winch. To the left of center is a stand fabricated from angle irons; the piece of equipment on top of the stand is the control valve for the hoist cylinders.

(**Opposite, above**) The right rear corner of the chassis of the M249 4 x 4 heavy gun-lifting truck is viewed. At the lower right is a hook on a bracket, welded to the rear bumper below the reflector; this was one of two hooks that served to hold the shackles at the bottoms of the hydraulic hoist cylinders of the lifting fork when the truck was not transporting a cannon. Above the rear suspension leaf springs are auxiliary springs; these were not present on the rear suspension of the M249 4 x 4 heavy gun-lifting truck that carried the rear of the atomic cannon.

(**Opposite, below**) More of the rear bumper of the M249 is seen from the right side. At the center is a tow pintle, and on the far end is the left hook for supporting the hydraulic hoist cylinder shackle when not in use. Towards the top of the bumper are two adjustable travel locks, to prevent the lifting fork from shifting from side to side during transit without the cannon and carriage. Under the bumper are the rear spring brackets. To the upper left is the right hoist cylinder.

U.S.A. 60146753

(**Above**) To the rear of the M249, the piston rods of the hydraulic hoists are extended down. Large shackles were pinned to the fittings at the bottoms of the pistons and these hooked onto the corners of the top carriage of the atomic cannon for purposes of raising and lowering the piece. The two fittings with square holes in them on the underside of the lifting fork, to the rear of the right hoist piston, were for locking the raised cannon carriage in place using a wedge. A similar lock is on the opposite side of the fork.

(**Opposite, above**) The lifting fork of the M249 is observed from the right rear. On the lifting forks of both the front and the rear truck, when the hydraulic hoist cylinders were attached to hooks on the corners of the top carriage of the cannon, and the cannon and the carriage were raised to the travel position, the carriage was secured in place with locking wedges. The flat undersides of the lifting forks formed a tight joint with the top of the carriage, making for a rigid structure for transport purposes.

(**Opposite, below**) On the underside of the lifting fork are the left brackets for the wedge lock of the cannon carriage, and the left piston of the hydraulic lift cylinder, with a fitting for a lift shackle at the bottom of it. The tires of the M249 were size 16.00-25, 20-ply. The axle track of the rear tires, measured from the centerlines of the dual wheels, was approximately 84 inches, while the track of the front tires was about 90.5 inches.

MILITARY

The lower part of the left hoist cylinder of an M249 and its piston are at the center of the photo. The upper part of the cylinder is in a well inside the fork; the round opening to the well is visible. To the left is the rear bumper of the M249, including the tow pintle. To the lower right is the front plate of the top carriage of an atomic cannon; on the lower right corner of the plate is the hook that engaged the shackle on the right lifting cylinder for hoisting and lowering the cannon and carriage.

(**Opposite, above**) The rear bumper of a different M249 is seen from the left side. On the upper part of the bumper are the two stops for the lifting fork. Retainer chains for the pins for locking the stops are on the outer sides of the bumper. To either side of the tow pintle are recessed tail light assemblies.

(**Opposite, below**) Taken from the left rear of an M249, to the left are the spare tire and a 5-gallon liquid container holder. Next to the tire are the high-pressure relief valve (left) and the low-pressure relief valve (right) of the hydraulic system. At the bottom and the right are the top of the fifth wheel and the front of the lifting fork. At the top, to the right of center, is the hoist cylinder control valve.

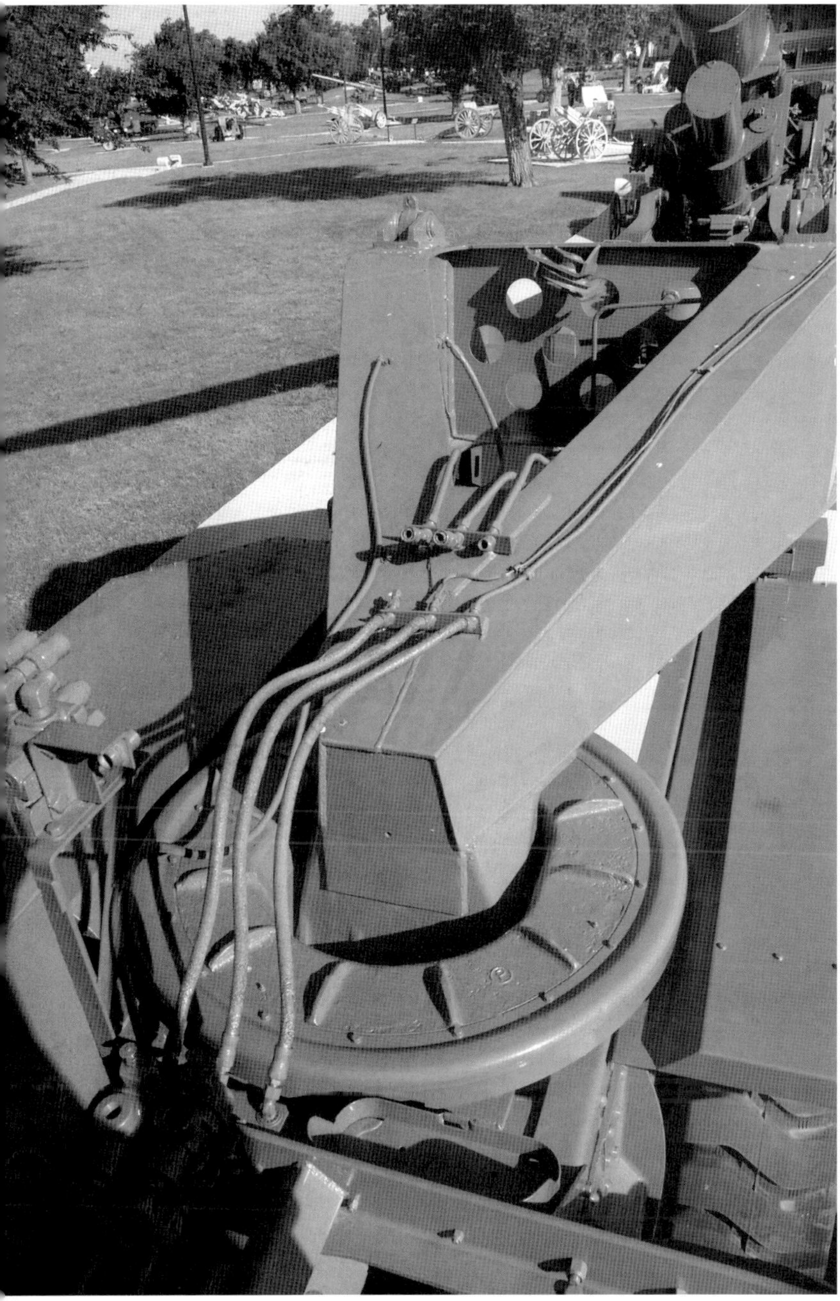

More of the fifth wheel and the lifting fork of the M249 is visible, from the same vantage point as the preceding photograph. At about the front, to rear, center of the lifting fork is a large cross member, perforated with lightening holes. Note the hydraulic lines and connections on the lifting fork. To the extreme left is the hoist cylinder control valve.

(**Opposite, above**) From the left side of the lifting fork, the left rear of the M249 is shown with the spare tire absent, enabling a clear view of the spare-tire carrier and davit. Also in view are the perforated muffler shield and part of the hydraulic reservoir. Between the front and rear left wheels is the fuel tank.

(**Opposite, below**) As seen from the area on the rear of the M249 where the spare tire would have been stored, to the left is the fifth wheel and the forward part of the lifting fork. To the right is the left rear walkway, situated above the left rear dual wheels.

The spare tire and the spare-tire davit on the M249 are seen from the left side of the vehicle, with the rear of the side of the engine compartment to the far left. The boom of the davit is resting against the spare tire; at the base of the boom or the top of the davit staff, is the hand-operated winch.

(**Opposite, above**) The entire left side of the engine compartment enclosure of the M249 is shown. The metal cover is removed from this side of the compartment, allowing light to enter it and enabling a view of the air cleaner, to the rear of the cab.

(**Opposite, below**) The entire top of the engine compartment of the M249 is shown. The two outer compartment covers are mounted on piano hinges on their inboard sides. Two grab handles and a recessed latch handle are towards the outboard sides of these covers. On the center panel is the fan shroud shield, with the fan inner shroud in the center of it. This shroud shield lacks the raised muffler shield added to the shroud on M249s around 1959. In the foreground are the muffler and its shield; in the background is the rear of the cab.

The left side of the engine compartment is viewed from the rear with the cover removed. The air cleaner is to the front. Below the right side of the opening is the oil cooler, mounted at an angle on the upper side of the engine. On the opposite side of the engine in the corresponding location is the transmission oil cooler.

(**Opposite, above**) On the left side of the cab of the M249, a slave receptacle is below the grab handle to the rear of the door. Grab handles also are on the door and the cab cover. To the front of the door are, top to bottom, a rear-view mirror bracket; blackout headlight and brush guard; service clearance light; and blackout clearance light.

(**Opposite, below**) A wheel on an M249 is shown close-up. It is mounted with twelve large hex nuts. The twelve smaller inner lock nuts are all fitted with a locking wire. The wheels were fitted with Bendix-Westinghouse air brakes.

U.S.A.
60146753

4A 867F

(**Opposite, above**) In a photo taken from the front of the left boarding ladder (far right) under the cab of the M249, the main focus is on the steering mechanisms. To the upper right is the pinion housing assembly, which takes directional power from the steering column in the cab above, and translates it, via the universal-joint shaft, to the steering-control housing at the center of the photo. From there, power is transmitted through the Pitman arm to the drag link, which operates the steering arm (lower right). Jutting from the near side of the front of the steering-control housing is the power cylinder.

(**Opposite, below**) On the right side of the front bumper are a tow eye and the front winch cable guide. The handle on top of the guide is attached to a roller shaft. By removing that shaft, the right-hand horizontal cable roller could be removed to make room to pull the winch cable loop through the guide. The hole on the outboard side of the guide aided in threading the winch cable and its loop through the bumper and into the guide.

(**Above**) Fort Sill preserves this example of an M250 4 x 4 heavy gun-lifting truck, the rear, pusher prime-mover of the Gun, Heavy, Motorized, 280mm, M65, or atomic cannon. The M250 was of significantly different design than the M249, having a rear-mounted cab, an engine compartment overhanging the rear of the vehicle, a front-mounted lifting fork and hydraulic hoists, and a lack of a winch and a spare tire.

(**Above**) The M250's front bumper is different from that of the M249, having the headlights mounted in it instead of on the cab front, and lacking the prominent hooks for the lift shackles. The travel locks on the upper part of the bumper are locked in place, in the travel configuration when not hauling the cannon, to keep the lifting fork from shifting from side to side.

(**Opposite, above**) The two travel locks for the lifting fork are of a swiveling design, with the bottom being secured to the bumper with an L-shaped pin with a retainer chain, and the middle being attached to a small swiveling bar on the bumper. This design allowed for securing the locks at several different angles, as desired. Note the curved rain guard over the service headlight.

(**Opposite, below**) Between the front and rear left wheels of the M250 is a sheet-metal mudguard for the steering control housing and related mechanisms, consisting of a bent front/bottom piece and a side in the shape of an irregular quadrilateral. A similar mudguard was not necessary on the M249 because its steering gear was to the front of the front left tire.

U.S.A. 60146754

Part of the steering gear is visible to the rear of the mudguard. The universal joint shaft from the steering column in the cab to the steering control housing behind the mudguard is visible above the toolbox to the front of the dual rear wheels. Below the angled bottom of the mudguard is the bottom of the Pitman arm and the drag link.

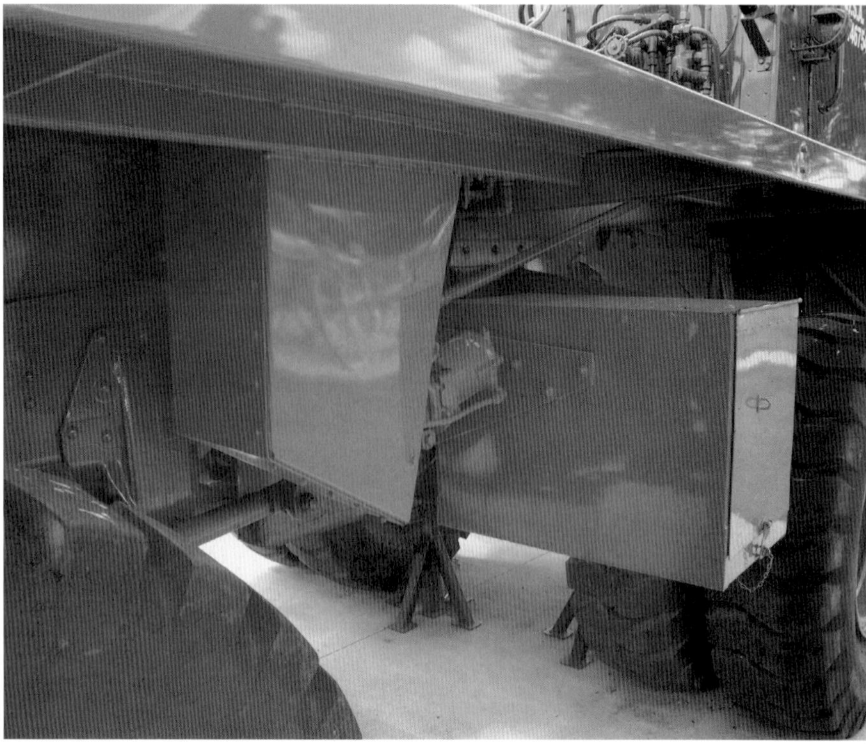

On the left side of the front of the cab is a maze of fittings and hoses for the hydraulic hoists on the lifting fork, including hoist cylinder control valve, a 30 psi check valve, and a 5 psi check valve. The hoist hydraulic pump and hydraulic reservoir are in the front end of the vehicle between the front wheels. To the right, on the left side of the cab, is the blackout headlight and brush guard.

The hydraulic hoses and valves for the hydraulic hoists are shown on the front of the cab of a different M250. These lines are routed through holes in an egg-shaped plate that is screwed onto the cab front. The two windshields of the M250 had top hinges and could be raised, but also the entire windshield assembly had a piano hinge along the bottom and could be lowered, and the bracket with the bumper on it, along with a similar one on the opposite side of the cab front, supported the windshield assembly when lowered.

An overall view of the left side of the M250 4 × 4 heavy gun-lifting truck provides a clear sense of the relative scales of the truck and the lifting fork. The fork on the M250 was longer than that of the M249, which hauled the front end of the atomic cannon. This was due to the fact that the fork had a long span to deal with from the fifth wheel, over the engine and hydraulic compartments, and forwards to the cannon's top carriage.

(**Above**) The boarding ladders on the M250 are to the rear of the rear tires. As seen on the left ladder, the bottom rung was removable and was attached by hex screws to the fixed part of the ladder. Next to the ladder is the left rear mudguard. Above are the cab and the engine compartment.

(**Opposite, above**) The rear bumper of the M250 is painted in yellow and black caution stripes. On the bumper are two towing eyes, tail-light assemblies, and a tow pintle. Above the bumper at the center of the rear of the engine compartment is an access door with a piano hinge at the top. To the left of the door is a holder for a 5-gallon liquid container.

(**Opposite, below**) Above the engine compartment is the rear of the cab, fitted with two windows. Jacks have been placed below the rear axle to relieve the springs of some of the weight of the vehicle. Each mudguard is made of two panels with a vertical joint between them, the inboard panels being narrower than the outboard ones.

4A 867F
A 14
U.S.A.
60146754

4A 867F
A 14
U.S.A.
60146754

(**Above**) The underside of the M250 is viewed from under the rear bumper. At the upper left are the three air reservoirs. To the upper rear of the rear axle is a light assembly. Details are available of the leaf springs, the extra-long U-bolts, and the left rear spring bracket.

(**Opposite, above**) The underside of the rear of the M250 is seen from a slightly different angle. At the top center is the floor of the engine compartment. Both of the rear spring brackets are in view. Note the absence of the auxiliary leaf springs above the springs; a feature found on the M249 4 × 4 truck that carried the front end of the atomic cannon top carriage.

(**Opposite, below**) On each end of the rear bumper of the M250 is a staff for a clearance flag, which helped the drivers gauge how close the vehicles were to objects to the sides. Four fasteners are on each of the staffs.

A 14
4A 867F
U.S.A.
60146754

A 14
U.S.A.
60146754

(**Opposite, above**) A fire-extinguisher holder is on the rear of the right running board. The round hole in the side of the engine compartment is for the intake of the air cleaner, the bulk of which is visible inside the rear grille. Above the front grille is a grab handle.

(**Opposite, below**) The right side of the engine compartment is seen from another angle, showing the grilles, the intake for the air cleaner, the service and blackout clearance lights, a fire-extinguisher holder, and the grab handle.

(**Above**) The woven-wire construction of the grilles on the right side of the engine compartment is depicted.

On the center panel of the engine-compartment cover is the fan shroud shield, protruding through the center of which is the muffler shield. Around 1959 this arrangement replaced the original configuration, which had a prominent muffler positioned on the middle of the fan shield, with a fan shroud extension around those components.

The right side of the cab of the M250 is viewed from the ground. Mounted on the cab below the windshield is a pioneer tool rack, which would have held an axe, shovel, and mattock. On the front corner of the cab is a bracket and bumper to hold the windshield assembly when lowered. The round plate below the tool rack towards the lower left is a holder for the bridge-classification number.

The engine compartment of the M250 juts far to the rear of the dual wheels, and even the rear of the cab is aft of the rear of the tires. Note the round access cover towards the bottom of the door.

(**Above**) The entire front of the cab of the M250 is displayed, showing the maze of hydraulic lines and valves that supplied power to the hoists. To the right is the fifth wheel and the rear end of the lifting fork. Below the fork and just above the level of the front deck is the propeller shaft from the engine to the transfer, which is located in the body approximately even with the rears of the front tires.

(**Opposite**) The front right corner of the cab of the M250 is viewed from above. On the side of the cowl is the bracket and the arm for a rear-view mirror. On the running board are two non-slip panels.

(**Below**) This is a view across the front deck of the M250 from the right side. At the center is the fifth wheel; at the upper right is the rear part of the lifting fork. To the left is the propeller shaft, with a universal joint, that transmits power from the engine behind the cab to the transfer in the forward part of the vehicle.

(**Above**) In the front deck of the M250 there is an open space up the middle, exposing to view some of the machinery inside the front end of the vehicle. As seen from the right side of the M250, to the left is the Allison TX-500 transmission. Above the transmission is the propeller shaft to the transfer, visible at the center of the photograph. The diagonal brace on the side of the transfer is the torque link.

(**Opposite, above**) The two windshield panels of the M250 have piano hinges at the tops, so they can be swung open for ventilation. Also the entire frame that holds the windshield panels has a piano hinge along the bottom edge, so the frame and windshields can be lowered. To do so, it was necessary to remove the cab top. Also visible are the two windshield-wiper arms.

(**Opposite, below**) The fifth wheel and the rear of the lifting fork of the M250 are to the right, and the front of the cab is to the left. Note the spokes cast into the top of the lifting fork; a casting number is visible on it. The propeller shaft for the transfer also is in view.

STOP

(**Opposite, above**) The hydraulic hoses and valves for the hydraulic hoists are shown on the front of the cab of a different M250. These lines are routed through holes in an egg-shaped plate that is screwed onto the cab front.

(**Opposite, below**) The spoked design on the top of the rear of the lifting fork of the M250 is seen from another angle. Weld beads are around the perimeter of it. Note the vertical indentation and the circular access plate on the center of the front of the cab.

(**Above**) The front end of the lifting fork of the M250 at Fort Sill is seen from the right side of the front deck of the vehicle. In the background is the rear of the atomic cannon. Routed on top of the fork are service and throttle air lines, and interphone line for communications between the two transporter trucks.

U.S.A.

(**Opposite, above**) Between the front and rear tires on the right side of the M250 is the fuel tank, supported by a strap hanger. On the deck above the tank is a round opening that allows access to the fuel filler neck. At the front of the fender, the rear of the right service headlight housing is visible.

(**Opposite, below**) Welded to the bottom surface of the lifting fork are two block-shaped structures that rest on the front bumper when the fork is in the lowered position. The right one is seen here adjacent to the right travel lock for the lifting fork. The brackets on the side of the fork are a non-original modification and appear to be for cable storage.

(**Above**) In a frontal view of the lifting fork and the front bumper of the M250, the fittings on the angle at the top of the branches of the fork are designated lifting brackets, and these provide trunnion bearings for the tops of the lifting hoists. Hanging below the fork are the pistons of the lifting hoists.

The example of a 280mm Gun M65, or atomic cannon, nicely restored and on display at the US Army Artillery Museum, Fort Sill, is observed from the left side, with the rear of the M249 4 × 4 heavy gun-lifting truck to the left. This cannon is the one nicknamed Atomic Annie that fired the only live nuclear round ever fired by a 280mm atomic cannon, during the Grable test in Operation Upshot-Knothole at the Nevada Test Site on 25 May 1953.

Nestled in the front end of the top carriage of the 280mm Gun M65 is the hydro-pneumatic equilibrator equipment, which acted to balance the mass of the gun in such manner that, if the electro-hydraulic elevation mechanism should fail, the heavy cannon could be elevated manually with ease. The square hole in the top corner of the side of the carriage is for inserting a steel wedge to secure the carriage to the lifting fork of the front transporter.

(**Opposite, above**) At the very front of the left side of the top carriage, a wedge and its fastening key are stored in brackets. A folding ladder is secured to the carriage; it was used for boarding the carriage when it was raised on the transporters. The plumbing is for compressed air for the throttle override service line and the brake override service line. Corresponding lines on the M249 would be coupled to these when the cannon was being transported.

(**Opposite, below**) Visible above the top carriage are the tops of several elements of the equilibrator; on the left is the oil reservoir and on the right are the two expansion tanks. This photo provides another view of the opening for the left front wedge, below which is the wedge, with its fastening key (right) and locking pin (left) flanking it. To lock the carriage to the lifting fork, the wedges were driven through the aligned openings in the carriage and the lifting fork. The fastening keys were inserted into the wedges from the top and the locking pins were inserted in holes in the bottoms of the locking keys to keep them from popping out.

(**Above left**) The horizontally oriented hydraulic cylinder of the equilibrator is viewed from the front left corner of the top carriage of the 280mm gun M65. On its inboard side are the two expansion tanks. To the far left is part of the oil reservoir of the equilibrator.

(**Above right**) On the right side of the expansion tanks is the right hydraulic cylinder of the equilibrator system. In the foreground is the cover of the air filter for the cylinder, with a perforated plate at the center that is fastened with four hex screws.

(**Above**) On the lower part of the cradle of the 280mm Gun M65 are three cylinders of the primary recoil system. The large one at the center is the recuperator cylinder, and the two smaller ones to the sides are recoil cylinders. The primary recoil system was the mechanism by which the gun tube recoiled and was recuperated within its cradle.

(**Opposite, above**) The main elements of the equilibrator are viewed from the front right corner of the top carriage, with the air cleaner to the left and the oil reservoir and the expansion tanks to the side of it. On the far side, the front of the hydraulic equilibrator cylinder is partially visible. Protruding from the top of the oil reservoir is an air filter; protruding from the bottom of the reservoir is the equilibrator motor.

(**Opposite, below**) The equilibrator motor is viewed from a different angle. The cylinder below the equilibrator motor is a nitrogen tank.

(**Opposite, above**) This view from below the right front corner of the top carriage shows, in the foreground, cylinders of the secondary recoil system (the mechanism whereby the entire top carriage recoiled and was recuperated independently of the cannon tube, when the cannon was fired). Also in view is the firing base of the bottom carriage.

(**Opposite, below**) Taken from a slightly different angle than the preceding photo, from left to right are the two cylinders of the secondary recoil system and the nitrogen tank of the equilibrator. The object to the top with hex nuts arrayed around the perimeter is the bottom of the oil reservoir of the equilibrator.

(**Above**) The front of the right side of the top carriage is shown, with the lifting fork of the M249 front transporter truck to the right background. To the rear of the boarding ladder are two dummy couplings with retainer chains. These were placed on air line couplings when not in use to keep out dirt and foreign matter.

(**Above**) On the top carriage of the right side of the atomic cannon at Fort Sill are, left to right: the equilibrator motor control switch and box; the elevating speed control handle and quadrant; equilibrator pressure gauge and relief valve; the elevating selector handle; and the elevating hand wheel. Below that hand wheel is the elevating brake pedal.

(**Opposite, above**) On the side of the right trunnion, above the elevating hand wheel, is an M1 quadrant. On top of the gun cradle is the right counter-recoil buffer. Below the gun tube is the right primary recoil cylinder and the primary recuperator cylinder.

(**Opposite, below**) The right-hand cradle lock (bottom right) and cradle-pin socket (on the side of the gun cradle above the cradle lock) are shown. When the gun is configured for travel, the cradle lock and the cradle-pin socket are in alignment and a pin is inserted through them to immobilize the gun's elevation.

(**Above**) The right cradle lock is seen close-up. Below the lock is a small placard that reads, 'WARNING / LOCK RECOIL BEFORE / RELEASING CRADLE / LOCKS'.

(**Opposite, above**) The forward floor of the top carriage of the atomic cannon is seen from the right side, with piston rods and linkages for the equilibrator system visible. To the lower left is the cradle elevating rack. In the background is the lifting fork and the rear end of the M249 4 x 4 truck.

(**Opposite, below**) This is the right side of the firing base of the atomic cannon, along with the right leveling jack and caster and the ribbed guides for the secondary recoil. The over/under pairs of hex screws of the recoil guide have locking wires. There are three jacks for cross-levelling purposes: one on each side and one to the rear. The hooks along the perimeter of the firing base are for attaching stabilizing anchors.

U.S.A. 60146753

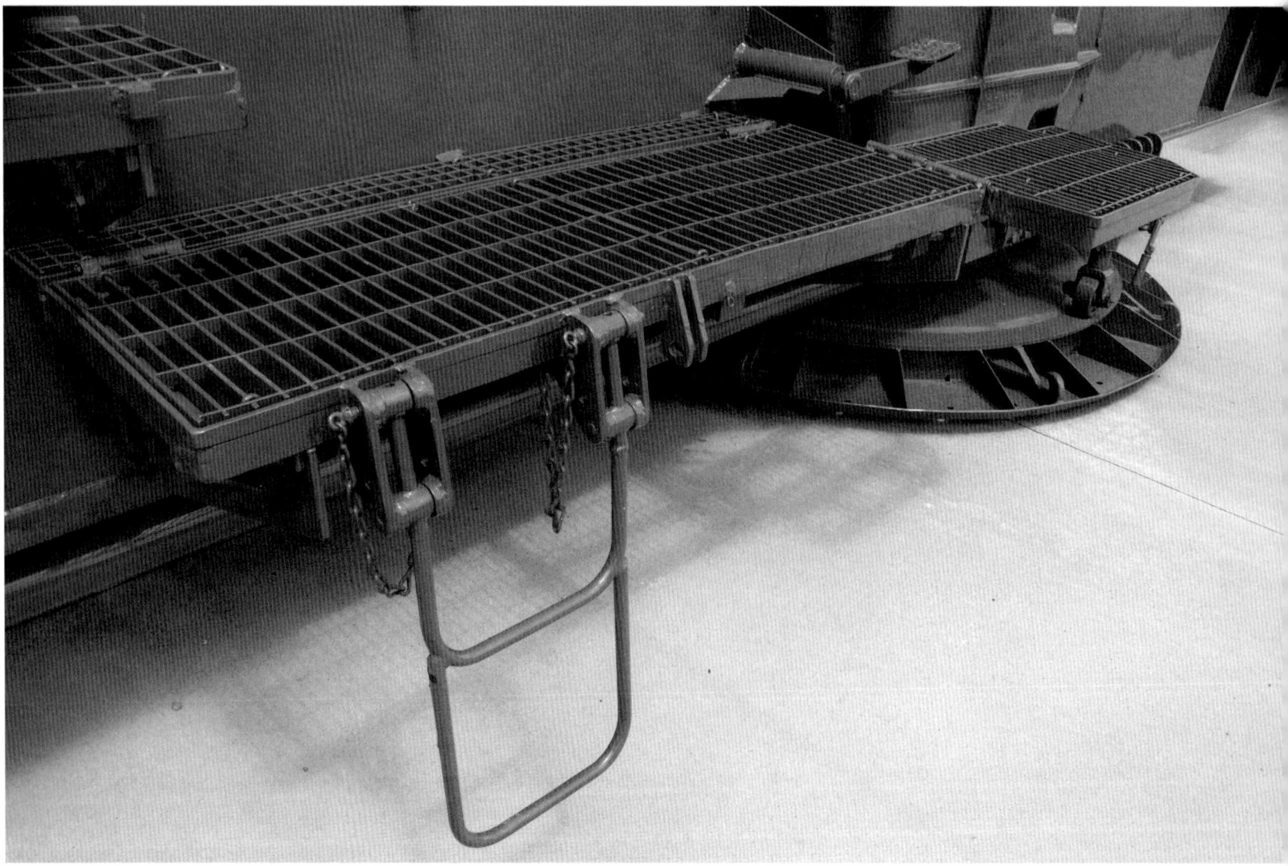

(**Opposite, above**) The right side of the firing base is seen from a different perspective, offering details of the elevating hand wheel, the cover for the elevating gears, and one of the folding platforms. Below the platform is the right tie rod, one of two connecting the firing base to the float underneath the rear of the top carriage.

(**Opposite, below**) On the long platform on the right side of the top carriage of the atomic cannon is a folding ladder. Early on, the ladder had a single rung. The second rung is a later modification. Pins with retainer chains secure the ladder in the 'down' position.

(**Above**) The controls on the right side of the atomic cannon's top carriage are displayed. On the trunnion is the M1 elevation quadrant, with a protractor scale to the rear (left in the photo) of it. Just aft of the elevation quadrant is the breech of the cannon. The elevating hand wheel is turned counter-clockwise to elevate the cannon and clockwise to depress it. The handle on the hand wheel could be folded by pulling a locking pin on its side.

The right trunnion of the atomic cannon is observed from the rear, providing a side view of the M1 elevation quadrant (right) attached to the side of the trunnion. At the bottom center, next to the platform, is the elevating brake pedal.

In a photograph that is a continuation of the preceding one, only pointing more downward, to the upper left is a protractor scale, associated with the M1 quadrant at the upper center. To the lower right is the elevating selector handle, to the front of which is the elevating hand wheel, seen in profile. The elevating speed control handle is at the lower left.

(**Opposite, above**) Details of the grille-type folding platforms on the right side of the top carriage of the atomic cannon are shown. Note how the small platform to the right is hinged on the larger platform to the left, which in turn is hinged on a narrow section of grille welded to the side of the carriage. At the center is the elevating brake pedal.

(**Opposite, below**) The small rear platform and the larger platform on the right side of the top carriage are observed from the rear. The platform hinges are leaf-type. Supporting the platforms are pivoting braces, which are folded against the side of the carriage when the platforms are folded up for travel.

(**Above**) The top of the 280mm gun cradle is viewed from the right side, showing the counter-recoil buffers and the gun-retraction sheave, which is a pulley wheel with a cable retainer used with the winch cable to move the gun barrel from travel position to firing position. The cable routing was from the winch, around the sheave and back to a fitting on the top front of the gun breech.

To the rear of the breech of the 280mm cannon (right) is a well between the sides of the top carriage, containing the loading platform, the rear floor, and the ramming mechanism. At the bottom are two sections of the folding trough on which the rammer pushes ammunition into the breech.

In this view of the breech of the 280mm cannon from the rear floor, in the center foreground is the rammer buffer, the part of the rammer that made contact with the rear of the projectile being pushed into the breech. The buffer rests in the fixed, rear part of the ammunition-loading trough. To the lower right and the lower left are the front ends of two retraction rails, tracks upon which two rollers on the bottom of the breech ride when the cannon tube is being retracted to its travel position.

The top carriage of the atomic cannon is viewed from the rear of the 280mm cannon breech to the rear. At the center of the photo is the rammer buffer seen in the preceding photo. It rests on the trough, with the folding continuation of the trough visible at the lower center. To the sides of the fixed part of the loading trough and extending towards the rear of the carriage are the retraction rails.

(**Above**) The fixed part of the ammunition trough (upper center of photo) is at the top of the ramming mechanism. Below the fixed trough is the housing for the rammer chain and the motor and gearing to drive the chain.

(**Opposite, above**) The rammer and rammer chain in the fixed part of the ammunition trough are seen from the right. Slightly below the front of the fixed trough is the right latch for the folding rammer trough; a left latch is on the opposite side. At the lower right is the right counterbalance spring of the folding trough.

(**Opposite, below**) The right trough latch is seen close-up at the top center of the photo. Below it is the right selector valve of the hydraulic system for the rammer.

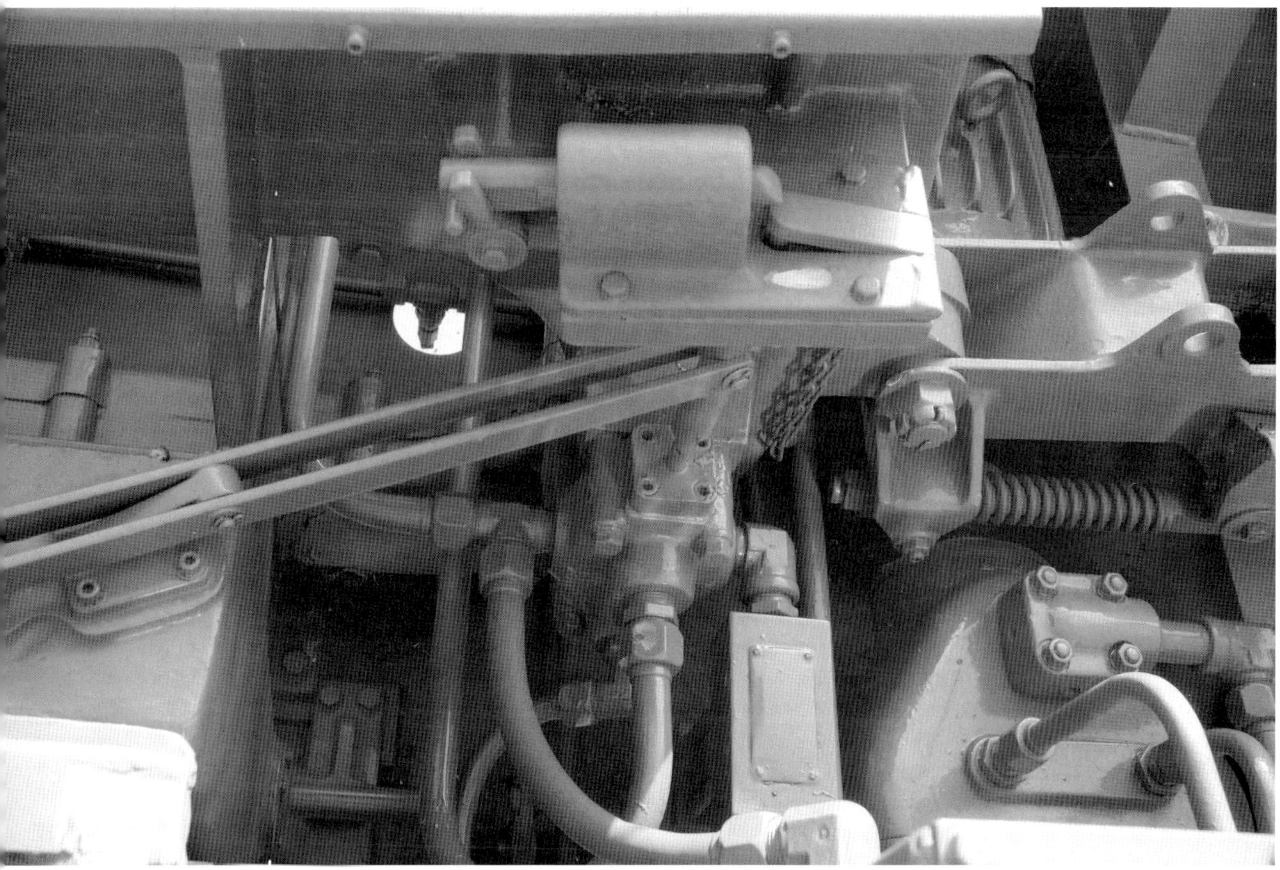

(**Opposite, above**) The rammer chain was not a continuous chain; rather, the end seen here was attached to the rammer buffer, pushing the buffer forwards to load a projectile or powder charge and pulling back to retract the buffer. To the rear of the visible portion of chain is the flange on the top rear of the ramming mechanism, which is bolted to the vertical plate. Below the flange is the rammer motor. Visible on the near side of this motor is a shaft with a universal shaft, used, along with a detachable hand crank, to manually operate the rammer in the event the motor failed.

(**Opposite, below**) The control lever for the rammer mechanism and a grab handle are in the foreground in this photo looking forwards on the right side of the atomic cannon. On the opposite side of the top carriage of the cannon is the loading davit. When the davit was rigged for lifting projectiles, the boom was supported at a higher angle by two stays made of steel cables.

(**Above**) The rammer speed-control handle is seen from the side. The placard behind the lever reads, 'DO NOT RAM / WITHOUT PROJECTILE.' In the background is the loading davit for hoisting projectiles from the ground to the rammer.

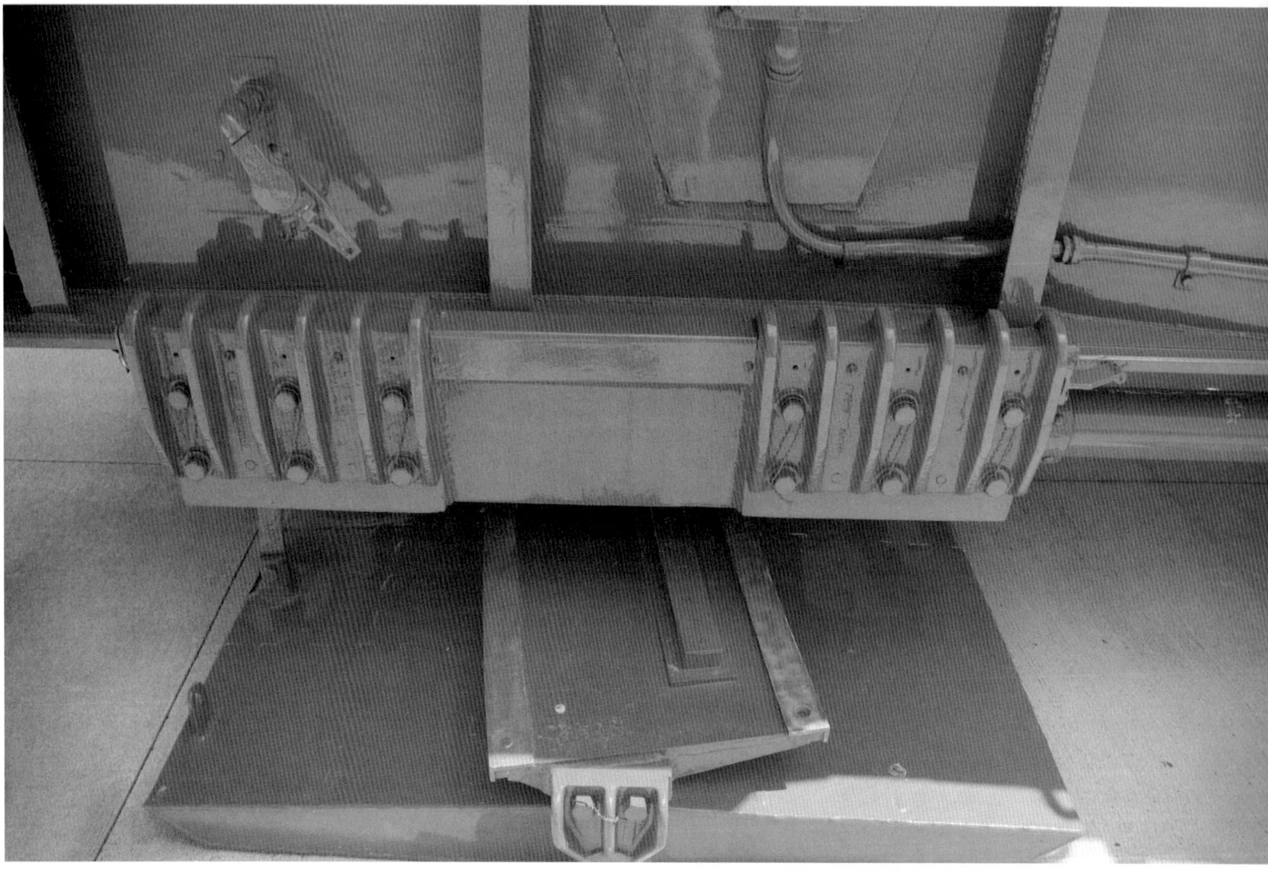

(**Opposite, above**) Below the rammer speed-control handle is a platform for the operator. A step is attached to the platform. Folding supports are on the two curved brackets. When the platform is folded up for travel, the two triangular lugs on the side of it are pinned to the swiveling bar on the side of the top carriage.

(**Opposite, below**) The float assembly that supports the rear of the top carriage and the cannon is seen from the right side. The ribbed guides for the secondary-recoil function are bolted to the float bolster, not visible here. At the bottom is the float pan, on top of which is the traversing rack and slide, the mechanism that allows for a total of 15 degrees of traverse for the cannon.

(**Above**) Jutting from the top sides of the top carriage are transporter arm sockets. These are fixtures that the two arms of the lifting fork fit into when attached to the atomic cannon for transport purposes. On the rear of the top carriage is the carriage rear door. Also visible at the rear corner of the top carriage is a lock wedge and its accessories.

The rear transporter arm sockets are seen from a different perspective. Similar transporter arm sockets are on the front end of the atomic cannon top carriage. At the bottom of the rear corners of this carriage are hooks to which shackles on the hydraulic lifts of the transporter vehicles were attached when the piece was to be transported.

The atomic cannon at Fort Sill is seen from above and to the rear. The bright-metal colored portion of the 280mm cannon barrel to the front of the gun cradle is part of the barrel's jacket and was milled and finished to a very high level of uniformity. This was so that that part of the jacket would constitute a bearing surface in contact with the inner surface of the gun cradle for purposes of recoil and recuperation.

(**Opposite, above**) As seen from the rear of the top carriage, the carriage upper door is made of diamond-tread plate, with a latch on the rear of it and two hinges to the front. Further forwards are the trough and the rammer buffer. In the foreground are the two gun-retraction stops.

(**Opposite, below**) The rear of the top carriage is observed from a different angle. When the gun was in full retraction for travel purposes, the brackets the two retraction wheels were mounted on under the breech fitted over the stops, and pins were inserted through holes in these brackets and stops to lock the gun barrel in place.

(**Above**) In a view of the left rear corner of the top carriage of the atomic cannon and the left side of the float below, the locking wedge and its fastening key and locking pin are at the center of the photo. Note the U-shaped swiveling handle on the locking wedge. Also present are air connections for the throttle and brake override service lines.

(**Opposite, above**) In another view of the rear part of the left side of the atomic cannon carriage, to the far left is the loading davit, with the swing handle, for steering the davit boom, folded up against the davit staff. Adjacent to the boom is the loading-davit winch housing, with the hand crank for the winch stored on top.

(**Opposite, below**) The float assembly, the rear element of the bottom carriage, is viewed from the left side. The tube to the left is the left tie rod, one of two connecting the front and the rear elements of the bottom carriage together. At the bottom are the float pan, the traversing rack, and the slide; at the rear is the float's travel lock.

(**Above**) The left side of the atomic cannon at Fort Sill from the loading davit winch housing forward is seen, with the platforms and steps deployed. The 280mm cannon is set at approximately the elevation for loading ammunition into it. It had an elevation range of 0 to 55 degrees. To the front of the cannon is the M249 4 x 4 heavy artillery transporter.

It took two crewmen to operate the hand crank for the loading davit; they did so standing on the ground. A crewman standing on the platform to the rear of the gun breech operated the swing handle, seen folded up against the davit staff, to swing the davit boom to the desired position.

(**Opposite, above**) The square lug in the hole near the bottom of the davit winch housing is for fitting the hand crank. The winch cable can be seen between the front of the winch housing and the the sheave at the bottom of the davit. The box to the rear (i.e., right) of the winch housing contains the AC magneto starter. Routed to the top of the magneto starter are a 2-conductor No. 14 wire cable and a 3-conductor No. 3 wire cable; routed to the bottom of the magneto starter is a 3-conductor No. 3 wire cable.

(**Opposite, below**) The support brackets for the loading davit are visible in this photo. Also apparent is the holder for the hand crank for the davit winch on the upper front facet of the winch housing. To the right is the M250 heavy artillery transporter and its front-mounted lifting fork.

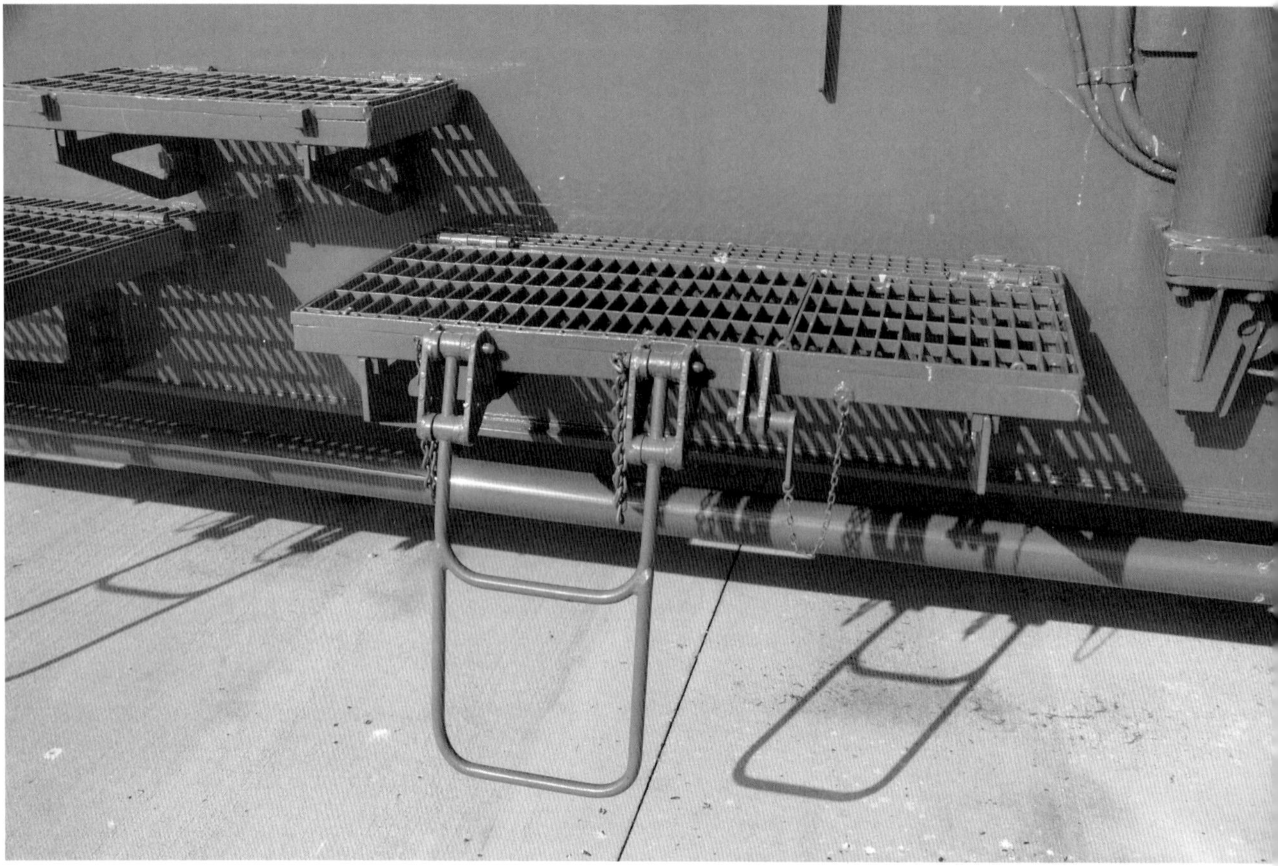

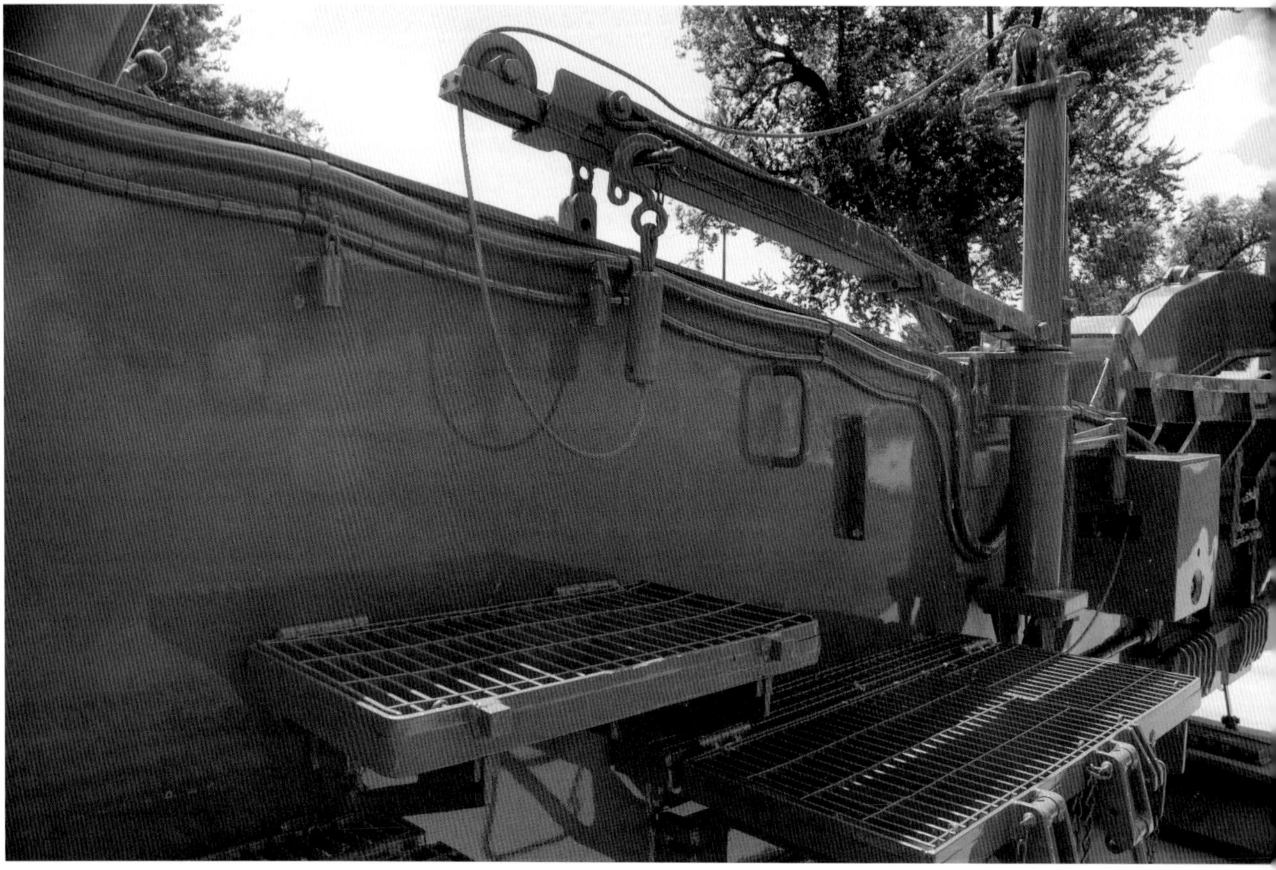

(**Opposite, above**) All three folding platforms on the left side of the atomic cannon's top carriage are in view. The platform with the steps lowered provided access to the loading platform and the loading mechanism to the rear of the cannon. It also was useful for crewmen when loading the cannon, allowing them a place to stand as they guided a projectile being hoisted up to the crewmen on the loading platform.

(**Opposite, below**) The hook at the end of the winch cable of the loading davit is hanging from a holder on the side of the boom. The cable lying along the boom is one of two stays that would be attached to the boom and to hooks on the top of the davit staff to strengthen the boom when it was in operation.

(**Above**) This is the view one obtains of the rear of the top carriage of the atomic cannon by standing on the platform to the immediate front of the loading davit. In the foreground is the rammer mechanism, to the sides of which are the retraction rails, and a good view is available of the design of the transporter arm sockets and the retraction stops.

The folding part of the loading trough is shown in the firing position, so it would be clear of the recoiling breech of the cannon. The fixed part of the loading trough is to the lower right. Pushing the handle on the side of the folding trough caused the trough to lower into the loading position.

(**Opposite, above**) The lower part of the right side of the folding part of the loading trough of an atomic cannon is to the far right. To the left is the breech of the cannon, which is set at a high angle of elevation, with mostly the top of the breech in view. Faintly visible below the breech is the loading platform, with diamond-tread surface and a cutout in the rear.

(**Opposite, below**) On the breechblock, the two horizontal handles with ball ends were for lifting the breechblock to close it. The breechblock was of an interrupted-screw design, so once the breechblock was closed, pushing the vertical handle forwards rotated the breechblock clockwise to lock it shut. On the top rear of the breech ring next to the lock handle is the eyebolt used with a winch cable to retract the cannon tube for travel.

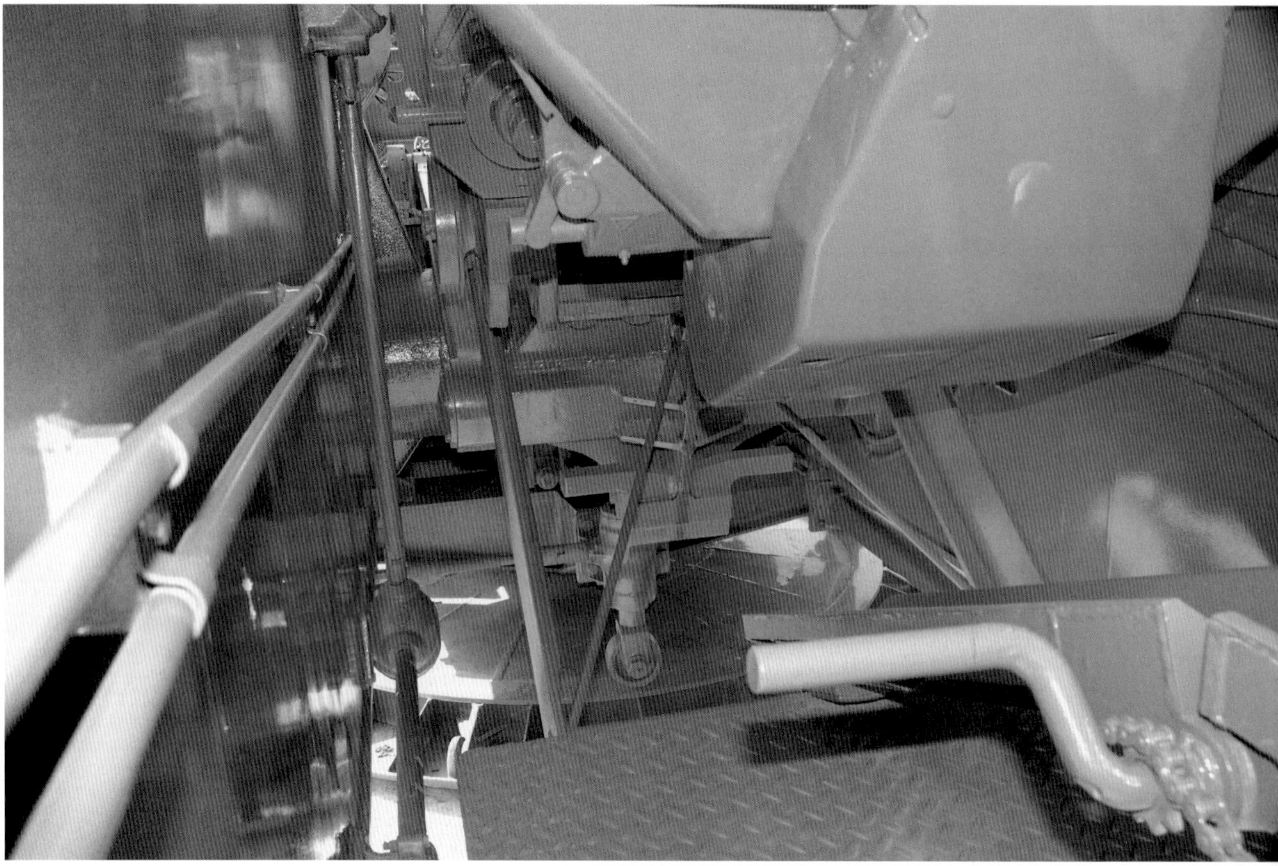

The 280mm cannon cradle and tube are viewed from the left side of the breech. The horizontally oriented gun retraction sheave is situated in a cutout in the steel plate that supports the counter-recoil buffers. To the lower left is the left trunnion.

(**Opposite, above**) In a view taken from the firing platform (bottom), the breechblock carrier and the lower part of the rear of the breech ring is at the top right. To the left of the breech ring is the left retraction wheel. Visible through the opening in the bottom of the carriage at the center of the photo is the firing base, part of the bottom carriage.

(**Opposite, below**) The left trunnion and the traversing hand wheel are shown. Mounted on the trunnion is the M30 telescope mount. The sighting equipment, an M12A7C panoramic telescope, would be attached to the telescope mount. To the rear of the hand wheel is a grab handle and a swiveling bar for securing a folding platform.

(**Above**) The left trunnion (center) is viewed from the rear from a different angle. Adjusting thumbscrews are on the bottom of the M30 telescope mount, and a socket is on top for mounting the M12A7C panoramic telescope.

(**Opposite, above**) In the top carriage of the atomic cannon below the recoil and recuperation cylinders at the top are the equilibrator expansion tanks, cylinders, and linkages. On the top edge of the carriage is the left cradle lock. In the background is the M249 front transporter truck.

(**Opposite, below**) The cradle elevating rack of the atomic cannon is viewed from the left front. A pinion at the end of the elevating gear train drove the cogs on the rack to elevate or depress the cannon. Attached to each side of the bottom of the rack is an arm from an equilibrator cylinder.

U.S.A. 6014675

This is the platform for the gun layer on the left side of the top carriage of the atomic cannon. The two-rung folding steps are deployed. On the edge of the platform to the right of the steps is a bracket which is pinned to the swiveling bar at the top right when the platform is folded up for travel.

(**Opposite, above**) All three platforms on the left side of the top carriage are lowered, in a view facing to the rear. The one in the foreground is for the gun layer. The steps are locked in the extended position with pins with retainer chains. In the background are the davit and the davit-winch housing.

(**Opposite, below**) On an unrestored atomic cannon, the left platforms are shown in the travel positions. The triangular supports for the platforms are folded inwards and locked in place. When the locking pins are removed from the top of the steps, the steps are allowed to swing down and be stored against the underside of the grille of the platform.

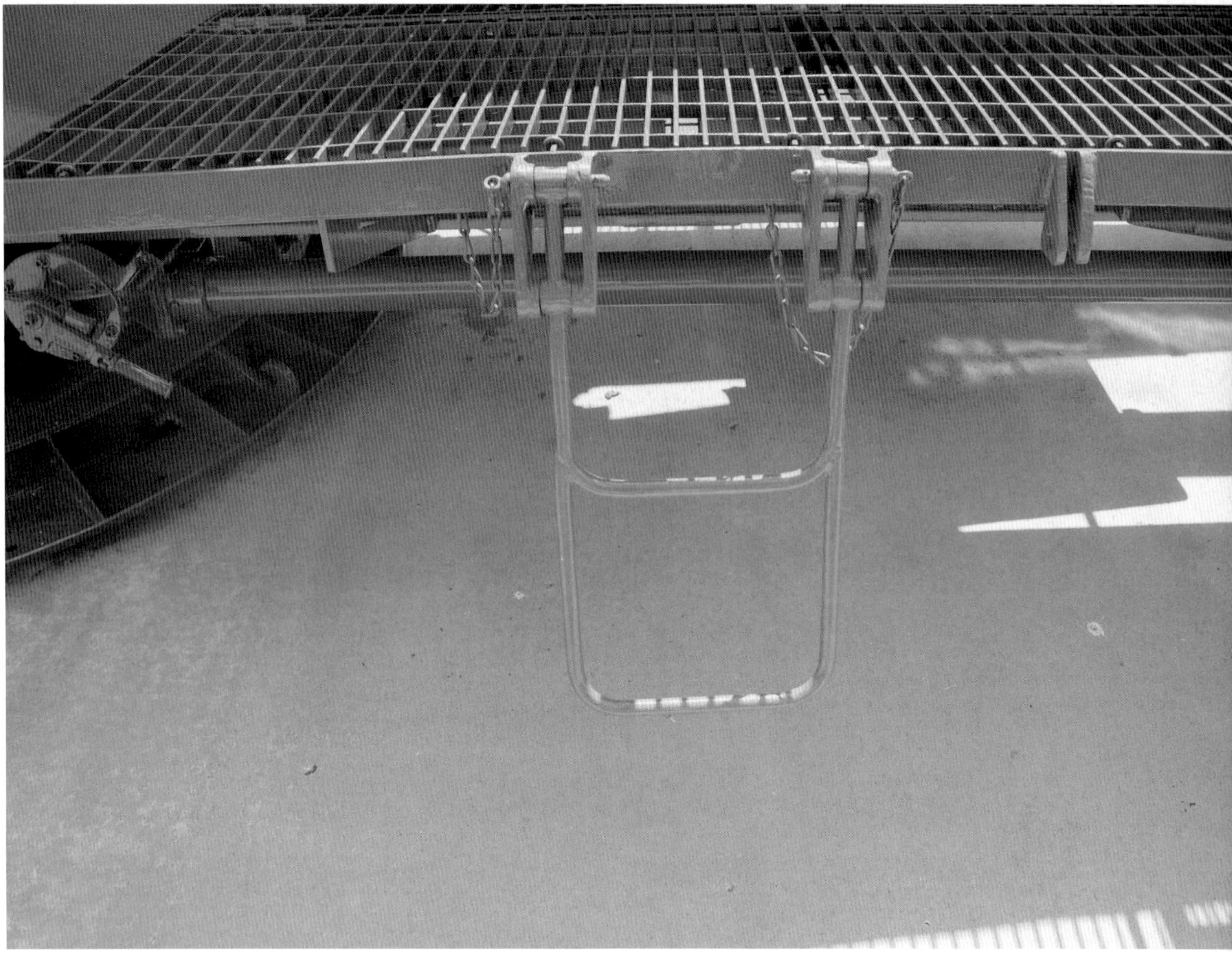

(**Above**) The steps on the front platform on the left side of the top carriage of the atomic cannon are seen from another angle, with the steps deployed. There are two sockets on each upright of the steps; the top one is for locking, and the bottom one supports the steps and allows them to pivot to the travel position. Note the device on the tie rod to the far left; this will be explained subsequently.

(**Opposite, above**) To the front of the forward platform on the left side of the top carriage is a manhole cover. The large box is the protective control box. Below and to the front of that feature are the power-circuit junction box and the receptacle box. Also in view is the left cross-leveling jack and caster.

(**Opposite, below**) In addition to the left and right cross-leveling jacks, for adjusting the levelness of the top carriage on the bottom carriage, there is a rear jack, situated underneath the carriage. This jack is operated by the hand crank shown here, mounted on the left tie rod. The hand crank's grip is of a folding design. Next to the handle of the crank is a socket for a metal bar, not installed. This hand crank transmitted force to the rear jack via a crankshaft.

(**Opposite, above**) The left cross-leveling jack is seen close-up. The caster of the jack is resting on a turntable rack on the top of the firing base. Note the hooks on the lower level of the firing base for attaching a stabilizing anchor. Also of interest are the two lubrication fittings on the top edge of the recoil guide to the right; these are located to either side of the vertical centerline of the guide.

(**Opposite, below**) Both of the front left recoil guides as well as the left cross leveling jack are shown. The forward guide has the same arrangement of lubrication fittings on top as the rear one. The L-shaped metal fitting extending from the caster yoke to the front of the jack's mounting plate is part of a limit indicator, which shows if the jack has met its raised or lowered limits.

(**Above**) Above the front left secondary-recoil guide are the receptacle box and, right, the power-circuit junction box. Projecting from the lower part of the forward side of the receptacle box is the receptacle and screw-on cap, for receiving power from an external generator. Above the receptacle box is a voltage-warning placard. Note the two mounting tabs and hex screws above the power-circuit junction box. The socket for a detachable bar on the left leveling jack is visible at the lower right.

Appendix One

General Data

280mm Gun Specifications

Caliber	280mm (11.02 inches)
Gun carriage length	84 feet, 2 inches
Firing position weight	94,000 pounds
Travel position weight	166,638 pounds
Weight of HE shell	605 pounds
Maximum rate of fire	1 round per minute
Muzzle velocity	3,500 feet per second
Range	31,200 yards (17.7 miles)
Breech block type	Stepped thread, interrupted screw
Firing mechanism	Percussion-electric
Minimum recoil	Primary, 32-inches; secondary 40-inches
Maximum recoil	Primary, 42-inches; secondary, 95-inches
Minimum elevation	0 mils
Maximum elevation	978 mils (55-degrees)

G-268 M249/M250 Prime Mover

Model	M249	M250
Weight (net)	37,293 pounds	35,341 pounds
Length	372.5 inches	373.5 inches
Width	123.5 inches	124 inches
Height	116 inches	103.75 inches
Tire Size	16.00 - 25	16.00 - 25
Max Speed	30 mph	30 mph
Fuel Capacity	140 gallons	140 gallons
Range Loaded	150 miles	150 miles
Electrical	24 Negative	24 Negative
Transmission Speeds	3	3
Aux Trans Speeds	3	3
Turning Radius	40 feet	40 feet

G-268 M249/M250 Prime Mover Engine Data

Engine Make/Model	Continental AO-895-4
Number of Cylinders	6
Displacement	895 cubic inches
Horsepower	375 @ 2,800
Torque	775 @ 2,100

Appendix Two

280mm Artillery Battalions, 25 June 1958

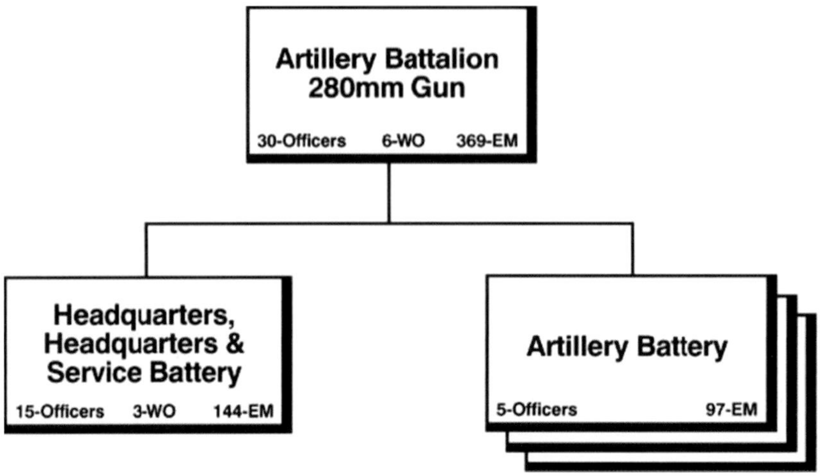

7th US Army, 42nd Artillery Group	
3rd Gun Battalion, 81st Artillery Regiment (280mm)	Baumholder, Germany
3rd Gun Battalion, 84th Artillery Regiment (280mm)	Baumholder, Germany
V Corps Artillery Group, 30th Artillery Group	
3rd Gun Battalion, 82nd Artillery Regiment (280mm)	Giessen, Germany
36th Artillery Group	
2nd Gun Battalion, 38th Artillery Regiment (280mm)	Darmstadt, Germany
72nd Artillery Group	
3rd Gun Battalion, 80th Artillery Regiment (280mm)	Bamberg, Germany
VII Corps Artillery	
3rd Gun Battalion, 39thArtillery Regiment (280mm)	Nürnberg, Germany
Headquarters and Headquarters Battery	
2nd Gun Battalion, 39th Artillery Regiment (280mm)	Fort Sill, OK
8th US Army	
3rd Gun Battalion, 38th Artillery Regiment (280mm)	Munsani, Korea

Appendix Three

Field Artillery Battery

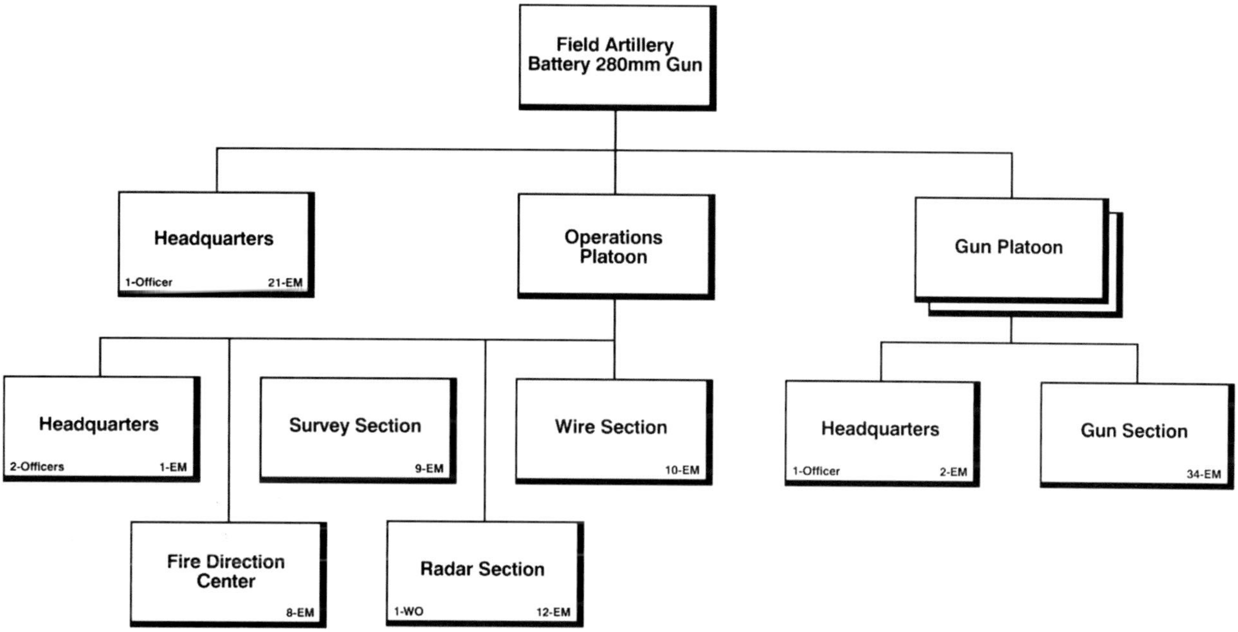

Operations Platoon Headquarters

Item	Qty
Truck Utility 0.25-ton 4 × 4 M38A1	2
Trailer Amphibious Cargo 0.25-ton 2-wheel M100	1

Battery Headquarters

Item	Qty
Tractor, Full Tracked Low Speed Diesel Driven Medium, D7C	1
Bulldozer blade, cable tilt for Tractor Medium	1
Trailer Cargo 0.75-ton 2-wheel M101	1
Trailer Tank Water 1.5-ton 2-wheel M149	1
Truck Cargo 0.25-ton 4 × 4 with winch M37	1
Truck Cargo 2.5-ton 6 × 6 LWB M35	2
Truck Cargo 2.5-ton 6 × 6 LWB with winch M35	1
Truck Tractor 5-ton 6 × 6 SWB with winch M52	1
Semitrailer Low Bed 15-ton 4-wheel M172	1
Truck Utility 0.25-ton 4 × 4 M38A1	2
Truck Van Shop 2.5-ton 6 × 6 M109	1

Survey Section

Truck Cargo 0.75-ton 4 × 4 M37 1
Truck Utility 0.25-ton 4 × 4 M38A1 2
Trailer Cargo 0.75-ton 2-wheel M101 1

Wire Section

Truck Cargo 0.25-ton 4 × 4 with winch M38A1 1

Fire Direction Center

Trailer Cargo 0.25-ton 2-wheel M101 1
Truck Cargo 0.75-ton 4 × 4 M37 1
Truck Utility 0.25-ton 4 × 4 M38A1 2

Radar Section

Trailer Amphibious Cargo 0.25-ton 2-wheel M100 1
Truck Cargo 2.5-ton 6 × 6 LWB M35 1
Truck Cargo 2.5-ton 6 × 6 LWB with winch M35 1
Truck Utility 0.25-ton 4 × 4 M38A1 1

Gun Platoon HQ

Trailer Cargo 0.75-ton 2-wheel M101 2
Truck Cargo 0.75-ton 4 × 4 M37 2
Truck Utility 0.25-ton 4 × 4 M38A1 2
Truck Utility 0.25-ton 4 × 4 with Radio Set AN/VRC-9 M38A1 2

Gun Sections

Gun Heavy Motorized 280mm M65 2
Truck Cargo 5-ton 6 × 6 LWB M54 4
Chassis Trailer 2.5-ton 2-wheel M200 w/30 KW Diesel Gen 2

Table of Organization and Equipment data derived from TOE 6-535D, 13 July 1960.